MOLYNEUX'S PROBLEM

ARCHIVES INTERNATIONALES D'HISTOIRE DES IDÉES

INTERNATIONAL ARCHIVES OF THE HISTORY OF IDEAS

147

MARJOLEIN DEGENAAR

MOLYNEUX'S PROBLEM

THREE CENTURIES OF DISCUSSION
ON THE PERCEPTION OF FORMS

MOLYNEUX'S PROBLEM

Three Centuries of Discussion on the Perception of Forms

MARJOLEIN DEGENAAR
Erasmus University Rotterdam

Translated from the Dutch by
Michael J. Collins

KLUWER ACADEMIC PUBLISHERS
DORDRECHT / BOSTON / LONDON

A C.I.P. Catalogue record for this book is available from the Library of Congress

ISBN 0-7923-3934-7

Published by Kluwer Academic Publishers,
P.O. Box 17, 3300 AA Dordrecht, The Netherlands.

Kluwer Academic Publishers incorporates
the publishing programmes of
D. Reidel, Martinus Nijhoff, Dr W. Junk and MTP Press.

Sold and distributed in the U.S.A. and Canada
by Kluwer Academic Publishers,
101 Philip Drive, Norwell, MA 02061, U.S.A.

In all other countries, sold and distributed
by Kluwer Academic Publishers Group,
P.O. Box 322, 3300 AH Dordrecht, The Netherlands.

Permission is granted to reproduce
Molyneux's letter to the authors of the *Bibliothèque universelle et historique*,
7 July 1688. Oxford, Bodleian Library, ms. Locke c. 16, fol. 92 *recto.*

Printed on acid-free paper

Printed in the Netherlands

To My Husband Gert-Jan
and
My Two Sons Sebastiaan and Diederik

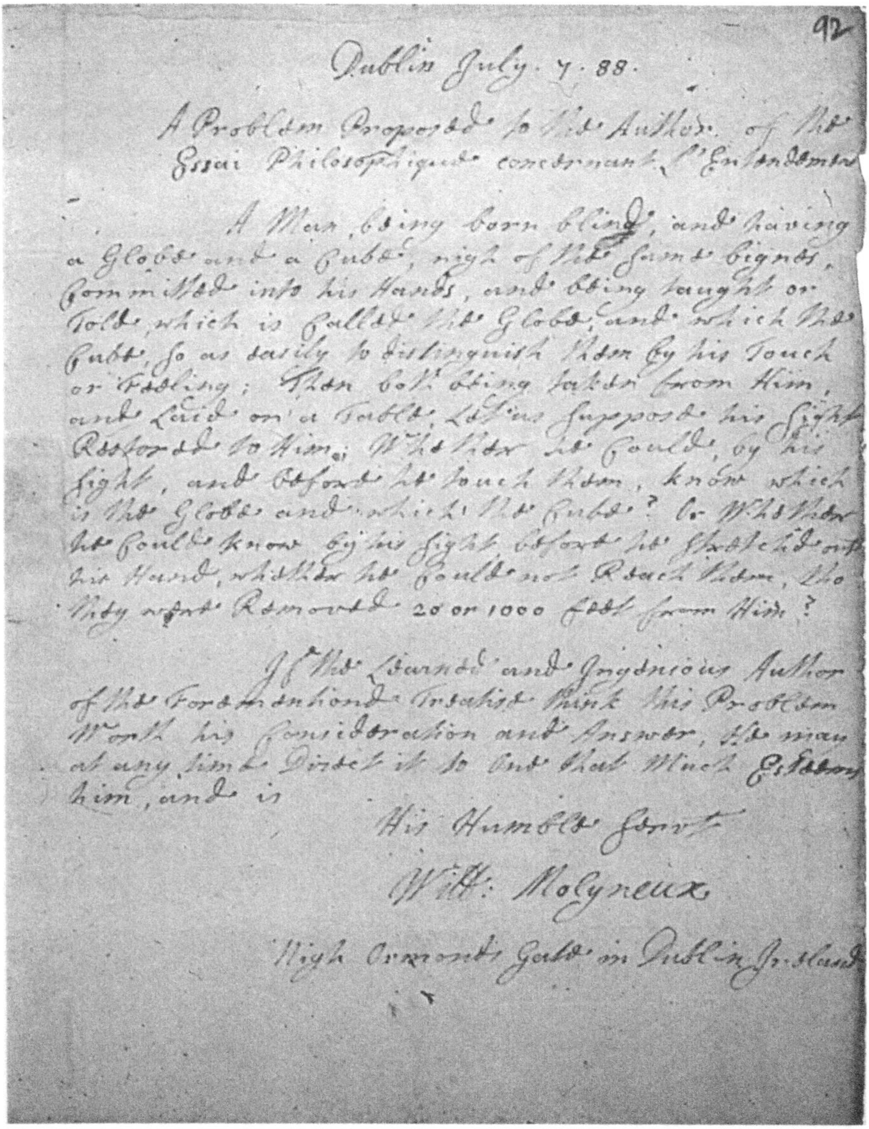

92

Dublin July. 7. 88.

A Problem Proposed to the Author of the Essai Philosophique concernant L'Entendement

A Man, being born blind, and having a Globe and a Cube, nigh of the same bignes, Committed into his Hands, and being taught or Told, which is Called the Globe, and which the Cube, so as easily to distinguish them by his Touch or Feeling; Then both being taken from Him, and Laid on a Table, Let us suppose his Sight Restored to Him; Whether he Could, by his Sight, and before he touch them, know which is the Globe and which the Cube? Or Whether he Could know by his Sight, before he stretch'd out his Hand, whether he Could not Reach them, tho they were Removed 20 or 1000 feet from Him?

If the Learned and Ingenious Author of the Forementioned Treatise think this Problem Worth his Consideration and Answer, He may at any time Direct it to One that Much Esteems him, and is

His Humble Servt

Will: Molyneux

High Ormonds Gate in Dublin Ireland

Molyneux's letter to the authors of the *Bibliothèque universelle et historique*, 7 July 1688. Oxford, Bodleian Library, ms. Locke c. 16, fol. 92 *recto*.

CONTENTS

PREFACE

ALTHOUGH there are already a considerable number of publications about Molyneux's problem concerning the visual capabilities of congenitally blind people who have just been made to see, the historical development of the discussion of the problem has never been studied in detail. This book aims to fill this lacuna. It contains a comprehensive survey of the discussion of Molyneux's problem from 1688 up to the present time.

Molyneux's question has been addressed by both philosophers and experimental psychologists. This study should therefore be of interest not only to historians of philosophy, but also to those interested in the history of experimental psychology, especially the history of vision research.

This book is based on the author's doctoral dissertation of 1992. She wishes to thank M. J. Petry and R. A. Crone for their stimulating criticism. She also wishes to thank her husband, Gert-Jan Lokhorst, for typesetting the book with TEX, and her two sons, Sebastiaan and Diederik, for providing her with welcome diversions from her academic work.

Chapter One

INTRODUCTION

IMAGINE that a congenitally blind person has learnt to distinguish and name a sphere and a cube by touch alone. Then imagine that this person suddenly recovers the faculty of sight. Will he be able to distinguish both objects by sight and to say which is the sphere and which the cube?

This problem, formulated in 1688 by the Irish philosopher William Molyneux, has intrigued a wide variety of intellectuals for three centuries. Those who have attempted to solve it include not only such philosophers as Locke, Berkeley, Reid, Leibniz, Voltaire, La Mettrie, Condillac and Diderot, but also such psychologists as Johannes Müller, Hermann Helmholtz and William James.

Molyneux's brainchild, which has gone down in history as "the Molyneux problem," was not something trivial, but rather a question closely linked with a number of major philosophical and psychological problems. It played an important part, for instance, in the development and justification of theories of perception and knowledge. It led to discussions on what can be directly perceived by the various senses, on the possible mutual interaction of the senses, on the influence of reason on perception, on the relationship between visual and tactile sensations of forms, on the relationship between visual and tactile concepts of forms and on the application of familiar concepts in new situations.[1] It constituted a central controversial question for eighteenth-century empiricists and rationalists, it figured in the nineteenth-century empiricism-nativism debate and it has also engaged the attention of researchers from a variety of disciplines in this century.

In a somewhat wider perspective, Molyneux's problem increased interest in the notion that our world undergoes far-reaching changes when we lack one or more of our senses, a suggestion found widely in the scientific and popular literature of the Enlightenment.[2] People lacking one or other of the senses were regarded as interesting from the point of view of theories of knowledge

---

[1] By visual and tactile forms I mean forms as perceived by the sense of sight or the sense of touch respectively. By visual and tactile concepts I mean concepts of forms obtained one way or another viâ sight or touch respectively.

[2] See, for example, Dufau 1837, Von Erhardt-Siebold 1932, MacLean 1936, Nicolson 1946, and Paulson 1987. See also Lende [1940] 1953.

because it was thought that they could serve to demonstrate what types of knowledge we possess thanks to the various distinct senses. From other angles, the blind, the deaf and the lame were also seen as curious creatures since—according to some—they were thought to possess not only another capacity for acquiring knowledge but also different beliefs, morals and æsthetics. In the mid-eighteenth century theoretical interest in the deaf and the blind was combined with a humanitarian interest which led to philanthropists such as l'abbé de L'Épée and Valentin Haüy being able to put through social reforms, including provision of care and education for deaf mutes and the blind.

Finally, the history of Molyneux's problem illustrates elegantly how an area of research traditionally included under philosophy was affected by numerous factors (including the use of new methods and techniques and the information they provided) and has become part of other disciplines such as psychology and neurophysiology.

The aim of this study is to investigate how Molyneux arrived at his problem, how the problem has been interpreted in the course of history, in what types of context it has been discussed and what sort of arguments have been evinced to justify the various solutions. At the same time an attempt will be made to isolate factors that have played a rôle in the changes in interpretation and argumentation.

The idea of sketching a history of the development of Molyneux's problem is not entirely new. Between 1772 and 1782 the Swiss philosopher Jean-Bernard Mérian published a number of articles in which he attempted to present a *histoire raisonnée* of Molyneux's problem. Of more recent date are the article by John Davis (1960), the book by Michael Morgan (1977) and Appendix A of John Gerald Simms's biography of Molyneux (1982). Obviously Mérian's writings are limited to the eighteenth century. Davis and Simms gave only a brief historical survey of the discussion about the problem. Morgan's somewhat disorganised book contains many digressions on matters which only indirectly relate to Molyneux's problem. It would thus seem desirable and justified to write a more adequate and more extensive history which would also cover the nineteenth and twentieth centuries.

This book is organized as follows. The *second chapter* gives a description of Molyneux's problem with a sketch of the background. The *third chapter* concentrates on the arguments originally used in the attempts to solve the problem. As will be seen, it was at first regarded as a purely hypothetical problem amenable to solution by philosophical analyses regarding the identity (or otherwise) of visual and tactile sensations (or concepts) of forms. The solutions proposed were usually closely connected with the positions taken up by their proponents in the empiricism-rationalism debate.

The discussion took a new turn in 1728 when the English surgeon William Cheselden published a report of the observations made by a patient, blind from birth, who had undergone a successful cataract operation. Philosophers believed that this provided them with empirical data that would be of help in answering Molyneux's question. However, all sorts of difficulties arose which stood in the way of a definitive answer. One of the consequences was that criteria were laid down governing satisfactory psychological experimentation. The *fourth chapter* studies the influence of Cheselden's report on the discussion of Molyneux's problem.

From the end of the eighteenth century a number of changes can be seen in the interpretation and treatment of Molyneux's problem, though there is no sense of a break with the past. Whereas in the eighteenth century the discussion centred principally on the relationship between visual and tactile sensations and concepts of forms, in the nineteenth century spatial vision stood in the centre of attention. The empirical data which were used in the discussion were of three kinds. In the first place, further cataract operations led to new results, some of which contradicted those recorded by Cheselden—which comes as no great surprise, since pre-operative and post-operative circumstances varied from case to case. Because the various cases were difficult to compare one to the other, none of the attempts to arrive at an unambiguous answer to Molyneux's question were successful. Secondly information concerning the visual capacities of newborn animals and babies came to widen the perspectives of the discussion. A number of researchers regarded the fact that some animals can see objects at a certain distance immediately after birth as proof of the notion that a person cured of congenital blindness would be able to distinguish a cube from a sphere. The question as to whether the same person would be able to *name* the objects was not, however, amenable to an answer in this way. Finally Wheatstone's discoveries in the field of spatial vision were related to Molyneux's problem, but again failed to provide a definitive solution to the problem. All this information was used in the nineteenth-century empiricism-nativism debate in which Molyneux's problem was frequently discussed. These nineteenth-century developments are dealt with in the *fifth chapter*.

Even though in the twentieth century Molyneux's problem has no longer played the major rôle it had in previous centuries, it has still been capable of arousing the interest of philosophers, psychologists, ophthalmologists, physiologists and other researchers. Though such interest is first and foremost of an historical nature new approaches can also be detected. Attempts have been made, for instance, to solve the problem using data derived from visual deprivation experiments with animals and from the use of sensory substitution systems. The *sixth chapter* discusses these modern developments.

Finally, all the interpretations, approaches and solutions put forward over the years are assessed as to their value and suggestions are made as to what the present-day importance of Molyneux's problem might be.

Chapter Two

MOLYNEUX'S PROBLEM

1 The First Formulation of Molyneux's Problem

On Saturday 7 July 1688[1] William Molyneux (1656–1698) wrote a letter to John Locke (1623–1704) setting out for the first time his problem concerning the person born blind:

Dublin July. 7. 88

A Problem Proposed to the Author of the Essai Philosophique concernant L'Entendement

A Man, being born blind, and having a Globe and a Cube, nigh of the same bignes, Committed into his Hands, and being taught or Told, which is Called the Globe, and which the Cube, so as easily to distinguish them by his Touch or Feeling; Then both being taken from Him, and Laid on a Table, Let us suppose his Sight Restored to Him; Whether he Could, by his Sight, and before he touch them, know which is the Globe and which the Cube? Or Whether he Could know by his Sight, before he stretchd out his Hand, whether he Could not Reach them, tho they were Removed 20 or 1000 feet from him?

If the Learned and Ingenious Author of the Forementiond Treatise think this problem Worth his Consideration and Answer, He may at any time Direct it to One That Much Esteems him, and is

His Humble Servant
William Molyneux
High Ormonds Gate in Dublin. Ireland[2]

---

[1] According to the Julian calendar, which remained in use in England until 1752.

[2] Molyneux writing to the authors of the *Bibliothèque universelle et historique*, 7 July 1688 (original spelling). Oxford, Bodleian Library, ms. Locke c. 16, fol. 92 *recto*; reprinted (with slight changes) in Locke 1976–1990, vol. 3 (1978), no. 1064. In 1688 Molyneux was not a personal friend of Locke and thus he addressed his letter to "Les Auteurs de la Bibliothèque," c/o Mr. Waesberg, bookseller in Amsterdam, who had published Locke's work in 1688.

What was it that had caused Molyneux to formulate this problem and why did he submit it for Locke's consideration?

2 Background to Molyneux's Problem

William Molyneux (17 April 1656–11 October 1698) was an administrator and politician who was well known in the Ireland of his time.[1] Molyneux was an enthusiastic devotee of the new learning. He was particularly attracted to optics: he confessed himself "much enamoured with optics, for in them there is such a mixture of physics and mathematics that renders this study very pleasing."[2] Molyneux's interest can be seen in his lectures and writings. His talks to the Dublin Philosophical Society, which he established as a sister institution of the Royal Society of London, covered such topics as the moon illusion and the problem of double vision. In his correspondence with the astronomer royal John Flamsteed he tackled problems not only in astronomy, ballistics and maritime matters but also in various branches of optics. In 1692 Molyneux published the *Dioptrica Nova*, the first substantial book in English on optics.

Molyneux was not alone among seventeenth-century intellectuals in his interest in optical problems and related questions. It was in this century that the science of optics advanced by leaps and bounds. Around 1600 the telescope and the microscope were invented, with all their attendant consequences. In 1604 Johannes Kepler demonstrated that the eye's crystalline body is not light-sensitive, but a lens. He also discovered that the images of objects formed on the retina by way of the lens are *reversed* and *flat*, a revelation that confronted him (and scientists after him) with the question of why we see objects *the right way up* and *at a distance.*[3] Further, Snellius and Descartes formulated the law of the refraction of light; Huygens discovered the wave form of light and Newton proved that light consists of a mixture of colours.[4]

---

[1] See W. Molyneux 1803, C. Molyneux 1820, Chillingworth 1946, Hoppen 1963, Dahl 1968, Hoppen 1970, Kelly 1979, Simms 1982, and Breathnach 1984. *The Case of Ireland's Being Bound by Acts of Parliament in England, Stated* (1698) is Molyneux's most important political publication.

[2] Molyneux to Flamsteed, 11 April 1682, quoted in Simms 1982, p. 61.

[3] See Kepler 1604. It had long been a well known fact that a person with two eyes can perceive depth, but after Kepler's discovery of the two-dimensional retinal image this fact became an object of disbelief because no-one could explain the phenomenon. In 1838 Charles Wheatstone demonstrated that it is possible to perceive depth when presented with two flat images.

[4] See Sabra 1967 for more information on these developments.

Molyneux's interest thus fitted in well with the dominant intellectual climate of the time. It could also be that the blindness that struck his wife, Lucy Domville, in the first year of their marriage contributed to his passion for theories of vision.[1] Possibly this tragedy was partly responsible for inspiring Molyneux to formulate his problem of the person blind from birth.

The direct cause of the formulation of Molyneux's problem must be looked for in the *Essai Philosophique concernant l'Entendement*, to which he referred in his letter. This *Essai* appeared in February 1688 in the *Bibliothèque universelle et historique* under the title "Extrait d'un livre anglois qui n'est pas encore publié, intitulé *Essai Philosophique concernant l'Entendement*, où l'on montre quelle est l'étenduë de nos connoissances certaines, & la manière dont nous y parvenons." It was a French abstract of Locke's *An Essay concerning Humane Understanding*, to be published in London in 1690.

Following in the footsteps of Aristotle,[2] Locke drew a distinction between ideas that can only be obtained through one of the senses and those that can be obtained through several senses.[3] According to this distinction we obtain ideas of colour only through the eye and of sound only through the ear. Locke advanced the notion that anyone lacking one of the senses will never possess the ideas belonging to that sense: "si quelcun a toûjours été destitué de l'un de ses sens, il n'aura jamais eu d'idée qui appartienne à ce sens. C'est ce qui paroît clairement par ceux qui sont nez sourds, ou aveugles."[4] Other ideas, however, including those of space and shape, would be obtained thanks to more than one sense: "Outre cela il y en a d'autres [idées] qui viennent à l'Esprit, par plus d'un sens, comme le mouvement, le repos, l'espace, les figures, qui nous viennent par la Vüe et par l'Attouchement."[5] The problem set out in Molyneux's letter dealt precisely with one of these "ideas," namely that of the shape of objects.

We can well imagine that Molyneux was not interested in the question as to whether a congenitally blind person who suddenly gains the faculty of sight could name the colour of a red or a blue object. In fact while blind he would never have seen colours and would not have been able to say whether

[1] William Molyneux and Lucy Domville married on 19 September 1678; a little over two months later—on 24 November—Lucy had a stroke which rapidly made her blind. See Simms 1982, pp. 20–22.

[2] See, for example, Aristotle's *De anima*, bk. II, ch. 6. See also Mackie 1976, pp. 28–32.

[3] The term "idea" was used by Locke and his contemporaries with different meanings. This sometimes leads to problems in the interpretation of fragments of text. It is not, for instance, always clear whether they are concerned with sensations or with concepts. Whenever I give expression to the opinions of Locke and other philosophers I do that in their own terms. See Chappell 1994 for a more elaborate discussion of Locke's usage of the term "idea."

[4] Locke 1688, p. 51.

[5] Locke 1688, p. 52.

what he was now seeing was red or blue unless someone had told him. But while blind he would certainly have learnt (by touch) the difference between a sphere and a cube. He would have a concept of shapes; he would know the characteristics of a spherical object and of a cube-shaped object. Quite possibly Molyneux thus regarded it as worthwhile to try to find out whether a person cured of congenital blindness would be capable of distinguishing and naming by sight two objects that he had previously learnt to distinguish and name by touch alone. It is probable that Molyneux presented his problem to Locke because the latter had expressed certain opinions concerning the ideas of the congenitally blind and the ideas of shapes that can be obtained both through touch and through sight. An additional factor was that Molyneux greatly admired Locke's work.

In the *Dioptrica Nova* (which was published in 1692, a couple of years *after* Molyneux's letter to Locke) there is a passage where Molyneux explicitly indicates the function he accorded to a person cured of congenital blindness. In his discussion of the question why we see objects the right way up while the retinal image is reversed, Molyneux wrote that the terms "right way up" and "inversed" are relative to "up" and "down" or to "farther" and "nigher" to the centre-point of the Earth.[1] Molyneux stated that we see things in their natural position because "the Mind takes no notice of what happens to the Rays in the Eye by Refraction or Decussation, but [...] the Mind does hunt back by means of each Pencil of Rays [...] to the Point from whence it comes, and is thereby directed strait thereto."[2] The fact that this natural power ensures that we see trees and suchlike the right way up can be confirmed by the presumed reaction of someone standing on his head: he would regard the trees that he sees as being the right way up and himself as being upside-down.[3] And this would be even more apparent in the reaction shown by a person blind from birth who suddenly recovers the faculty of sight:

> This will be more evident to us, by considering the case of an adult Person, who has been blind from his Birth, and now suddenly restored to his Sight: He is not prejudiced by custom, and yet (doubtless) would judge as is usual.[4]

[1] Molyneux 1692, p. 105.
[2] Molyneux 1692, pp. 289–290.
[3] Molyneux 1692, p. 212.
[4] Molyneux 1692, p. 212.

3 The Publication of Molyneux's Problem

For reasons unknown Locke never replied to the letter Molyneux addressed to him on 7 July 1688. However, a couple of years later, after the two philosophers had started an amicable correspondence, Molyneux returned to his problem, this time with success. The exchange of letters was a result of Molyneux's praise of Locke in his dedication of the *Dioptrica Nova* to the Royal Society:

> But to none do we owe for a greater Advancement in this Part of Philosophy [i.e., Logick], than to the incomparable Mr Locke, who in his *Essay concerning Humane Understanding*, has rectified more received Mistakes and delivered more profound Truths, established on Experience and Observation, for the Direction of Man's mind in the Prosecution of Knowledge [...] than are to be met with in all the Volumes of the Antients. He has clearly overthrown all those Metaphysical Whymsies, which infected mens Brains with a Spice of Madness, wherby they *feign'd a Knowledge where they had none, by making a noise with Sounds, without clear and distinct Significations.*[1]

Locke was sent a copy of the *Dioptrica Nova*, read Molyneux's flattering words and thanked the writer: "if my trifle could possibly be an occasion of vanity to me; you have done most to make it so, since I could scarce forbear to applaud my self upon such a testimony from one, who so well understands demonstration."[2] He added: "Sir, you have made great advances of friendship towards me, and you see they are not lost upon me."[3] There followed an exchange of letters mainly dealing with improvements to Locke's *Essay*.

In one of these, dated 2 March 1692/3[4], Molyneux once again presented Locke with his problem of the person born blind, though in a somewhat altered form, asking Locke if he could perhaps find some place in his *Essay* to say something about it.[5] This time Locke reacted with enthusiasm: "Your ingenious problem will deserve to be published to the world."[6] From the second edition of his *Essay* (that of 1694) Locke included Molyneux's problem in his work and thereby made it accessible to a wide audience:[7]

---

[1] Molyneux 1692, "Dedication to the Royal Society."

[2] Locke to Molyneux, 16 July 1692, in Locke 1976–1990, vol. 4 (1979), no. 1515.

[3] Locke to Molyneux, 16 July 1692, in Locke 1976–1990, vol. 4 (1979), no. 1515.

[4] In England the new year began legally on 25 March. Like many others, Molyneux wrote double years during the period from 1 January to 25 March.

[5] Molyneux to Locke, 2 March 1692/3, in Locke 1976–1990, vol. 4 (1979), no. 1609. The letter is included in Locke 1708.

[6] Locke to Molyneux, 28 March 1693, in Locke 1976–1990, vol. 4 (1979), no. 1620.

[7] Because Molyneux's problem became known through Locke's *Essay* it is Locke's version that is quoted here. The edition from which it is taken is Nidditch's (1975), which is based

> *Suppose a Man born blind, and now adult, and taught by his touch to distinguish between a Cube, and a Sphere of the same metal, and nighly of the same bigness, so as to tell, when he felt one and t'other, which is the Cube, which the Sphere. Suppose then the Cube and Sphere placed on a Table, and the Blind Man to be made to see. Quære, Whether by his sight, before he touch'd them, he could now distinguish, and tell, which is the Globe, which the Cube.*[1]

The reader will have noticed that in this second version of his problem Molyneux omitted the question about distance. Perhaps he believed that the answer to the question regarding the distinguishing and naming of a sphere and a cube implied the answer to the distance question. It is also possible that he made the omission because in the *Dioptrica Nova* he had—clearly, he believed—demonstrated that distance cannot be seen and that the answer was therefore obvious.[2]

In the *Dioptrica Nova* Molyneux had stated that the estimation of distance was more a function of our judgmental capacity than of our vision, and that it was not something inborn but rather a capacity acquired through experience and comparison:

> In Plain Vision the Estimate we make of the *Distance* of Objects (especially when so far removed, that the Interval between our two Eyes, bears no sensible Proportion thereto; or when look'd upon with one Eye only) is rather the Act of our *Judgment*, than of *Sense*; and acquired by *Exercise* and a Faculty of *comparing*, rather than *Natural*.[3]

The reason Molyneux gave for this was later to play a fundamental rôle, especially in Berkeley's theory of vision:

> For *Distance* of it self, is not to be perceived; for 'tis a Line (or a Length) presented to our Eye with its End towards us, which must therefore be only a *Point*, and that is *Invisible*.[4]

---

on the fourth edition of 1700; for reasons of chronology quotations are given as "Locke [1694] 1975."

[1] Locke [1694] 1975, bk. II, ch. ix, §8. Instead of "of the same metal" Molyneux had written "(Suppose) of Ivory." See Molyneux to Locke, 2 March 1692/3, in Locke 1976–1990, vol. 4 (1979), no. 1609.

[2] Lievers has written that "the Molyneux problem is not primarily concerned with the application of concepts nor with the correlation between sight and touch, but must be placed within the context of a solution to the problem of depth perception" (Lievers 1992, p. 415). This may be true of the first formulation but hardly seems to apply to the second.

[3] Molyneux 1692, p. 113.

[4] Molyneux 1692, p. 113.

Molyneux stated that distance is mainly perceived by means of interjacent bodies—such as mountains, trees and houses—and by the estimation we make of the comparative magnitude of bodies and of the clarity of their colours. The distance from nearby objects, however, he believed to be observed by the turning of the eyes or viâ the angle of the optical axes.[1] A person born blind and seeing for the first time would, according to Molyneux's theory, have as yet no experience of such judgments and would thus be unable to estimate distance.

4 Conclusions

A number of plausible reasons can be found to explain why Molyneux formulated his problem and why it was Locke to whom he submitted it. In the first place Molyneux was interested in blindness because his wife was blind. Furthermore he involved himself in questions regarding optics and the psychology of seeing. In this he was a true child of his time, for the science of optics was flourishing in the seventeenth century, spurring on such great scientists as Kepler, Huygens and Newton to study a wide variety of geometrical, physical, physiological and psychological aspects of light and vision and subsequently making discoveries that appealed to the general interest. The rise of empirical philosophy also ensured that there was an increase in epistemological interest in the senses and was partly the reason why Molyneux's problem fell on fertile ground. Molyneux brought his problem to Locke's attention because the latter had elaborated in his *Essai* certain notions concerning the ideas of persons born blind and ideas of shape that could be acquired both by sight and by touch.

Where did Locke discuss Molyneux's problem and how did he attempt to solve it? What did Molyneux believe to be the correct solution? And what were the arguments advanced by subsequent philosophers to justify the positions they adopted?

[1] Molyneux 1692, p. 113.

Chapter Three

PHILOSOPHICAL DISCUSSIONS IN THE EIGHTEENTH CENTURY

1 Molyneux's Problem as a Hypothetical Problem

SINCE philosophers originally believed that it was impossible to restore the sight of a person born blind, they regarded Molyneux's problem as a kind of thought experiment, susceptible of solution by reasoning alone.[1] The arguments they gave for their solutions usually concerned the relationship between visual and tactile sensations or were based on the relationship between visual and tactile concepts of objects.

Everyone agreed that there is a difference between the visual and the tactile *sensations* of objects. However, there was no agreement about the *relationship* between the visual and the tactile sensations of a particular object. Some philosophers were of the opinion that the link between the two is arbitrary and is only forged by experience. Others believed that there certainly is a necessary relationship between the two. Some believed that this could be perceived immediately, while others regarded experience as necessary.

As regards the relationship between visual and tactile *concepts* of objects, here too there were differences of opinion. Some philosophers suggested that the visual concept of a sphere, for instance, differs from the tactile concept of a sphere, and that both concepts could be made to relate to one another either by experience or by understanding. But others believed that visual and tactile concepts of objects are essentially the same or at least have something in common, and that this could be immediately perceived.

Below we examine the way in which the answers to Molyneux's question related to the various opinions.

[1] Similar thought experiments are still being proposed today. Think, for example, of Frank Jackson's imaginary scientist Mary, who knows all about the neurophysiology of colour vision but has spent her life in a black-and-white environment. Would she learn anything she did not know before if she left her rooms? (Jackson 1982.) The analogy between Jackson's experiment and Molyneux's problem is noted in Levin 1986.

2 Negative Answers to Molyneux's Question

Molyneux assumed that the person born blind had learnt both to distinguish a sphere from a cube by touch and to give them their correct names. The question he put to Locke was actually a double question: if the man were to recover his power of sight would he be able to *distinguish* them by sight, without the help of the sense of touch, and would he be able to *name* them? Molyneux himself did not give separate answers to these two questions, for his answer was as follows:

> *Not. For though he has obtain'd the experience of, how a Globe, how a Cube affects his touch; yet he has not yet attained the Experience, that what affects his touch so or so, must affect his sight so or so; Or that a protuberant angle in the Cube, that pressed his hand unequally, shall appear to his eye as it does in the Cube.*[1]

Experience will have taught the man that a cube has projecting angles which exert uneven pressure on the hand, while a sphere feels the same over its entire surface. But he will as yet not have had the experience of the impressions made by these angles and by this regular object on the eye. The man would not know that what he was seeing were a sphere and a cube because he would not have had the experience that what he was seeing had anything to do with what he had previously felt. In brief, Molyneux believed that the relationship between tactile and visual sensations of the shape of objects would not be immediately apparent but would have to be learnt. And it is possible that he regarded such a relationship as necessary.

In his *Essay* Locke expressed agreement with Molyneux's statement, but he did not seem particularly interested in the link between the sense of touch and the faculty of sight: in the chapter dealing with perception he used Molyneux's problem to illustrate the thesis that we often have false beliefs about the way in which we perceive—without our taking notice of it. Locke stated that when, for instance, we look at a uniformly coloured sphere, the idea we get of it is that of a (flat) circle with a variety of shades and colours. But from experience we have learnt that this sort of idea is caused by a sphere and thus we interpret the idea of the unevenly coloured circle as the idea of a uniformly coloured sphere. This happens so quickly that we hardly notice it.

In order to make this process clearer, Locke used a language metaphor which was to take on a persistent life of its own in discussions on Molyneux's problem. Whenever a person reads or listens to something with attention

---

[1] Locke [1694] 1975, bk. II, ch. ix, §8. See also Molyneux to Locke, 2 March 1692/3, in Locke 1976–1990, vol. 4 (1979), no. 1609.

and understanding, stated Locke, he does not take note of the letters or the sounds but of the concepts that they call forth.[1] Just as sounds are *signs* of concepts, so also is a circle a *sign* for a sphere. The passage in which Locke put his opinion regarding unconscious judgments reads as follows:

> We are farther to consider concerning Perception, that the *Ideas we receive by sensation, are often* in grown People *alter'd by the Judgment*, without our taking notice of it. When we set before our Eyes a round Globe, of any uniform colour, *v.g.* Gold, Alabaster, or Jet, 'tis certain that the *Idea* thereby imprinted in our Mind, is of a flat Circle variously shadow'd, with several degrees of Light and Brightness coming to our Eyes. But we having by use been accustomed to perceive, what kind of appearance convex Bodies are wont to make in us; what alterations are made in the reflections of Light, by the difference of the sensible Figures of Bodies, the Judgment presently, by an habitual custom, alters the Appearances into their Causes: So that from that, which truly is variety of shadow or colour, collecting the Figure, it makes it pass for a mark of Figure, and frames to it self the perception of a convex Figure, and an uniform Colour; when the *Idea* we receive from thence, is only a Plain variously colour'd, as is evident in Painting.[2]

It is at this point that Locke introduced Molyneux's problem: "To which purpose I shall here insert a Problem of that very Ingenious and Studious promoter of real Knowledge, the Learned and Worthy Mr. *Molineux*."[3] Locke apparently expected that his statement could be tested with the help of a person born blind who has recovered the power of sight. Such a person would not be prejudiced by habit as are all normally sighted adults. Locke responded to Molyneux's question saying that he believed that a person born blind would not be able to say immediately with any certainty which was the sphere and which the cube:

> I agree with this thinking Gent. whom I am proud to call my Friend, in his answer to this his Problem; and am of the opinion, that the Blind Man, at first sight, would not be able with certainty to say, which was the Globe, which the Cube, whilst he only saw them: though he could unerringly name them by his touch, and certainly distinguish them by the difference of their Figures felt. This I have set down, and leave with my Reader, as an occasion for him to consider, how much he

[1] Locke [1694] 1975, bk. II, ch. ix, §9.
[2] Locke [1694] 1975, bk. II, ch. ix, §8.
[3] Locke [1694] 1975, bk. II, ch. ix, §8.

> may be beholding to experience, improvement, and acquired notions, where he thinks, he has not the least use of, or help from them: And the rather, because this observing Gent. farther adds, that *having upon the occasion of my Book, proposed this to divers very ingenious Men, he hardly ever met with one, that at first gave the answer to it, which he thinks true, till by hearing his reasons they were convinced.*[1]

What is noticeable here is that Locke did not deal with Molyneux's question regarding the ability of the newly-sighted individual to distinguish between the sphere and the cube but only treated the question of *naming* the objects. A further point worth noting is that Locke believed that the man would be *unable to say with certainty* which was the sphere and which the cube. We know that Locke assumed that we are able naturally only to see shapes in two dimensions.[2] A man born blind who gains his sight and who is not yet affected by habit would, in Locke's view, obtain from a sphere the idea of a circle and from a cube that of a square or some sort of hexagon, depending on his angle of vision.[3] If we assume that the man born blind were to be given the two objects to view separately, according to Locke's opinion he should at least be able to perceive a *difference* between the two objects, that is he should be able to distinguish them one from the other.[4] If he were to be presented with the two objects simultaneously (and in all likelihood that was Molyneux's idea), it would be somewhat more difficult. Perhaps he might suspect that the square or the hexagon had something to do with a cube and the circle with a sphere, but he would not know it for sure. In order to be able to name the cube and the sphere correctly he would first have to learn that certain two-dimensional projections of the objects that he is observing correspond in some way to tangible three-dimensional objects.[5] This could explain why Locke wrote that the man would not be able to say *with certainty* which was the sphere and which the cube.

---

[1] Locke [1694] 1975, bk. II, ch. ix, §8. Molyneux responded to Locke's writing with the following words: "My most Honour'd Friend, For so you have publickly allowd me to call you; and tis a Title wherein I boast more than in Maces or Parliament-Robes. [...] I can only Pour out my thanks to you for the Favourable Character under which you have transmitted me to posterity." Molyneux to Locke, 28 July 1694, in Locke 1976–1990, vol. 5 (1979), no. 1763.

[2] This conflicts with Locke's statement that we can gain a notion of space through the sense of sight. See Locke [1694] 1975, bk. II, ch. v, and bk. II, ch. ix, §8. See also Berman 1974b.

[3] See Park 1969 about "real" and "apparent" dimensions of objects. According to Bolton Brandt 1994, the man cured of blindness would not at first receive visual ideas of the bodies he looks at, but would only be aware of light and colour. See also Vienne 1992.

[4] According to Mackie 1976, p. 32, "Locke could [...] hold that we get the same idea of shape from both sight and touch, provided that this is confined to two-dimensional shape."

[5] See Brandt 1975.

Like Molyneux, Locke probably assumed that visual and tactile ideas of shape have an essential relationship to one another which can be learnt by experience; nowhere did Locke defend the idea of the complete heterogeneity of sight and touch, an idea which the Irish philosopher George Berkeley was most certainly to support, and his notions on this point are crystal clear.

While Locke used Molyneux's problem merely as an illustration of his thesis that we owe more to experience than we tend to admit and that unconscious judgments can play a part in perception, George Berkeley (1685–1753) gave it a central rôle in his philosophy. Ernst Cassirer even stated that "die *Neue Theorie des Sehens*, die den Auftakt zu Berkeleys Philosophie bildet und die alle ihre Ergebnisse implizite enthält, [...] nichts als der Versuch einer vollständigen systematischen Entwicklung und Aufhellung des Molyneuxschen Problems [ist]."[1] Because of the importance of Molyneux's problem for Berkeley's philosophy on the one hand and, on the other, the influence of Berkeley's theory of sight on the discussion surrounding Molyneux's problem, we need here to make a detailed examination of Berkeley's ideas.

One of his most important motivations was the fight against scepticism, the principal cause of which Berkeley believed to be the general belief in the existence of material entities or external objects. He made it his business, therefore, to overthrow this belief. Berkeley was interested in theories of vision because the existence of the visible world is usually—and, according to him, unjustifiably—used as an argument for the existence of a material world. He studied the works on optics written by Descartes, Barrow and Molyneux as well as Newton's *Opticks* which had just been published.[2] These thinkers studied mainly geometrical optics, whereas Berkeley was interested in the psychology of seeing, as witness his *An Essay towards a New Theory of Vision* (1709).

The aim of the *Essay* was "to shew the manner wherein we perceive by sight the distance, magnitude, and situation of objects. Also to consider the difference there is betwixt the ideas of sight and touch, and whether there be any idea common to both senses."[3] As Berkeley remarked in the Appendix to the second edition of his *Essay* in 1710, his whole theory rested on the theory that we perceive *distance* neither directly nor by means of anything else that has an essential link with it, such as lines and angles. Following in Molyneux's footsteps Berkeley stated:

---

[1] Cassirer 1932, p. 145. See also Teape 1870, p. 3: "[Berkeley's] philosophy can alone be truly known, when seen germinating from the question of Molyneux." Quoted by Luce [1934] 1967, p. 34, note 1.

[2] Descartes 1637, Barrow 1674, Molyneux 1692, Newton 1704.

[3] Berkeley [1709] 1975, §1.

> It is, I think, agreed by all that distance, of itself and immediately, cannot be seen. For distance being a line directed end-wise to the eye, it projects only one point in the fund of the eye, which point remains invariably the same, whether the distance be longer or shorter.[1]

Any idea not directly perceived, said Berkeley, must be perceived by means of another idea.[2] Ideas which, according to Berkeley, suggest distance include the sensation produced by the turn of the eyes (convergence), the unclarity of the appearance and the tension of the eye (accommodation). Among other factors contributing to the creation of the idea of distance he mentioned the quantity, the size and the nature of the objects we perceive. He believed that experience links such factors to the notion of distance.[3]

A person born blind gaining the power of sight would in the beginning, said Berkeley, obtain no idea of distance through the faculty of sight; the sun and the stars, nearby and distant objects would all appear to be in his eye—or, rather, in his mind.

> The objects intromitted by his sight would seem to him (as in truth they are) no other than a new set of thoughts or sensations, each whereof is as near to him as the perceptions of pain or pleasure, or the most inward passions of his soul. For our judging objects perceived by sight to be at any distance, or without the mind, is [...] entirely the effect of experience, which one in those circumstances could not yet have attained to.[4]

Berkeley stated that it is only after we have had long experience of certain ideas derived from the sense of touch being linked to certain ideas derived from sight that we can immediately conclude which tactile ideas will follow certain visual ideas in a natural way. Visual ideas *suggest* tactile ideas, just as a flushed face suggests embarrassment and a pale face suggests fear.[5]

In order to prevent confusion Berkeley distinguished two kinds of objects of sight: primary and secondary. By means of primary, direct objects of sight (viz. light and colours) we are, he said, able to perceive secondary, indirect

---

[1] Berkeley [1709] 1975, §2. The remark "It is [...] agreed by all" should probably be regarded as rhetorical. In fact Berkeley believed that it was the general opinion that we see things at a distance outside ourselves.

[2] Berkeley [1709] 1975, §9: "It is evident that when the mind perceives any idea, not immediately and of itself, it must be by means of some other idea."

[3] Berkeley [1709] 1975, §§16–28.

[4] Berkeley [1709] 1975, §41.

[5] Berkeley [1709] 1975, §45 and §65. In the *Philosophical Commentaries* Berkeley speaks of "the constant & long association of ideas." Berkeley [1707–1708] 1975, no. 225.

and improper objects of sight. Berkeley believed these latter properly to belong to the sense of touch.[1] Just as Locke had understood flat surfaces as signs of three-dimensional shapes and had compared the visual perception of three-dimensional shapes with the understanding of linguistic tokens, so Berkeley compared the perception of secondary objects of sight with the understanding of the meaning of words. Whenever we hear a familiar language we receive simultaneously the sounds and the corresponding meaning of the words. These two are, through experience, so tightly bound together that it seems as if we are hearing not sounds but meanings.[2] Where Locke made no more than a passing reference to the analogy between visual perception and the understanding of language, this analogy plays an important part in Berkeley's writings.

Berkeley believed that *size* as well as distance was perceived indirectly. Consequently he regarded visual size not as a primary but as a secondary or improper object of sight. The visual size of an object can change, whereas the tactile size is always the same. The estimates we make of visual size, said Berkeley, depend entirely on experience. An assessment of size made by a person born blind opening his eyes for the first time would be totally different from the assessment we would make.[3]

Berkeley introduced a third variant of Molyneux's problem relative to the perception of the *position* of objects. He thereby made clear the function he accorded to such a thought experiment. Berkeley believed that it is useful to place ourselves in the blind man's shoes in order to rid ourselves of our visual experience and its accompanying prejudices regarding visual perception. He was convinced that this would, to a certain extent, be possible:

> In order to disentangle our minds from whatever prejudices we may entertain with relation to the subject in hand [i.e., situation], nothing seems more apposite than the taking into our thoughts the case of one born blind, and afterwards, when grown up, made to see. And though, perhaps, it may not be an easy task to divest ourselves entirely of the experience received from sight, so as to be able to put our thoughts exactly in the posture of such a one's, we must, nevertheless, as far as possible, endeavour to frame true conceptions of what might reasonably be supposed to pass in his mind.[4]

The person born blind, said Berkeley, would not at first think that what he was seeing was high or low or the right way up or upside-down. It would only

[1] Berkeley [1709] 1975, §50.
[2] Berkeley [1709] 1975, §51.
[3] Berkeley [1709] 1975, §79.
[4] Berkeley [1709] 1975, §92.

be after some experience had been gained that he would learn that objects pictured on the lower part of his eye are above, since he would see them clearly by aiming his gaze upwards. Without this eye movement, terms related to the position of tangible objects, such as "right way up" and "upside-down," would never have been transferred to the ideas which belong to sight.[1] Berkeley believed that what had largely contributed to writers on optics being misled in this matter was that they had placed too much emphasis on retinal images. Whenever they thought of these tiny images they imagined that they were looking at the bottom of the eye of another person and that they saw the images which were pictured there.[2] In addition, Berkeley was of the opinion that it was erroneous to imagine that retinal images were representations of external objects, for in fact there would be nothing in common between ideas of sight and those of touch; moreover the direct objects of sight do not, he stated, exist without the mind.

This brings us to the second aim of the *Essay* regarding the heterogeneity of sight and touch. Berkeley ventured to make the following statement:

> *The extension, figures, and motions perceived by sight are specifically distinct from the ideas of touch called by the same names, nor is there any such thing as one idea or kind of idea common to both senses.*[3]

One of the arguments used by Berkeley in an attempt to support this statement was related to a new variant on Molyneux's question: a person born blind gaining his sight would not think that the objects perceived were of the same nature as the objects of touch nor that they had anything in common.[4] Another argument concerned the statement that only quantities of the same sort can be added together. In view of the fact that Berkeley could not imagine that he could add a visible and a tangible line together, he concluded that they were heterogeneous.

Finally Berkeley borrowed a further argument from Molyneux's problem.[5] If a square surface perceived by touch were to be of the same kind as a square surface perceived by sight, he stated, the person born blind would recognise it immediately on seeing it. It would be no more than "introducing into his

---

[1] Berkeley [1709] 1975, §98.

[2] Berkeley [1709] 1975, §116. See also Turbayne 1955.

[3] Berkeley [1709] 1975, §127.

[4] Berkeley [1709] 1975, §128. Once again Berkeley pointed out the usefulness of an experiment of this type: "And surely, the judgment of such an unprejudiced person is more to be relied on in this case, than the sentiments of the generality of men: who in this, as in almost everything else, suffer themselves to be guided by custom, and the erroneous suggestions of prejudice, rather than reason and sedate reflexion."

[5] Berkeley [1709] 1975, §132: "A farther confirmation of our tenet may be drawn from the solution of Mr. Molyneux's problem."

mind by a new inlet an idea he has been already well acquainted with."[1] We should therefore either assume that visual forms differ from tactile forms or that the solution proposed by Molyneux and Locke ("those two thoughtful and ingenious men") is incorrect. Berkeley pointed to his earlier statement that a person born blind gaining his sight would have names for objects previously perceived by touch, but would be unable to give these same names to objects perceived for the first time by sight:

> Cube, sphere, table are words he has known applied to things perceivable by touch, but to things perfectly intangible he never knew them applied. [...] the ideas of sight are all new perceptions, to which there be no names annexed in his mind: he cannot therefore understand what is said to him concerning them: and to ask of the two bodies he saw placed on the table, which was the sphere, which the cube? were to him a question downright bantering and unintelligible.[2]

Berkeley emphasised that it was erroneous to think that one and the same thing affected both the sense of touch and the sense of sight. If the same angle or shape which was the object of touch was also the object of sight, what would prevent the man from knowing it at first sight?[3]

In order to provide an answer to the question of why visual and tactile shapes are given the same name while not being of the same kind, Berkeley once again made a comparison with language. Words are not considered as entities in themselves, he said, but as signs of things. Since it would be superfluous to have a name for a thing and for a sign of that thing, it is customary to indicate both by the same name. The same would apply to forms: visual forms are signs of tactile forms and are hardly regarded as entities in themselves. It would be superfluous to have different names for the tactile form and the visual form which indicates it. This does not imply that tactile and visual shapes are of the same sort. A tangible square and a visible square are just as different as a tangible square and the six-letter word "square" which indicates it.[4]

Responding to the objection that a tactile square is more like a visual square than a visual circle, Berkeley stated that while this is true,

> it is not because it is liker, or more of a species with it, but because the visible square contains in it several distinct parts, whereby to

[1] Berkeley [1709] 1975, §133.

[2] Berkeley [1709] 1975, §135.

[3] Berkeley [1709] 1975, §136: "For though the manner wherein it affects the sight be different from that wherein it affected his touch, yet, there being beside this manner or circumstance, which is new and unknown, the angle or figure, which is old and known, he cannot choose but discern it."

[4] Berkeley [1709] 1975, §140.

> mark the several distinct corresponding parts of a tangible square, whereas the visible circle doth not. The square perceived by touch hath four distinct, equal sides, so also hath it four distinct angles. It is therefore necessary that the visible figure which shall be most proper to mark it contain four distinct equal parts corresponding to the four sides of a tangible square, as likewise four other distinct and equal parts whereby to denote the four equal angles of the tangible square.[1]

Just as certain visual forms are more appropriate to represent certain tactile forms, Berkeley stated, so also do certain written words better describe sounds. The fact that letters represent sounds is in itself arbitrary. But once it has become the object of common consent, the combination of letters representing a certain sound is no longer arbitrary.

The question could be put as to why visual and tactile ideas can be so easily confused, which is not the case with other symbols. Berkeley believed that this was so because visual signs are *constant* and *universal* and because their relationship to tactile ideas is learnt immediately after birth.[2] The signs are not determined by man but the objects of sight constitute "the universal language of Nature."[3] It is a universal language which instructs us in how to behave in order to acquire those things which are necessary for our existence and welfare, and in order to avoid that which can cause us harm. However, there is no necessary connection between visual and tactile figures and, said Berkeley, this becomes evident if we consider that what seems round and smooth to the touch may, when viewed under the microscope, appear quite different.[4]

Berkeley closed his *Essay* with a few remarks on *geometry*. On the basis of what we have already seen, he determined that the object of geometry is constituted not by visual but by *tactile extension and form*. An intelligent being having the sense of sight but lacking that of touch (an *unbodied spirit*, a kind of inverse of Molyneux's person born blind) would, he thought, be incapable of perceiving solid bodies or flat shapes.[5]

To round off this section we are going to look at a philosopher influenced by Berkeley (among others) but whose ideas regarding the relationship between touch and sight were somewhat more subtle: Thomas Reid (1710–1796), the founder of the Scottish school of common sense philosophy and later known for his faculty psychology. In his *An Inquiry into the Human Mind* (1764)

---

[1] Berkeley [1709] 1975, §142.
[2] Berkeley [1709] 1975, §144.
[3] Berkeley [1709] 1975, §147.
[4] Berkeley [1709] 1975, §105.
[5] Berkeley [1709] 1975, §§150–159.

Reid proposed two solutions to Molyneux's problem, one for when the man was using only his sight and one for when he was also capable of mathematical reasoning.

Reid had originally espoused the doctrine which stated that ideas are the only immediate objects of awareness but, like Kant at a later date, he was awakened from his slumbers by Hume's *Enquiry concerning Human Understanding* (1748). Reid considered Hume's philosophy to be a *reductio ad absurdum* of scepticism; it was a system that pulled the carpet from under our feet because it made it impossible to justify any of our beliefs. Reid came to the conclusion that the origin of scepticism was to be found in the doctrine of ideas, this latter being based on prejudice and confusion. Reid was thus of the opinion that the notion that the immediate data of perception are ideas was nothing more than a hypothesis which could nowhere find support. Moreover, the notion failed to fulfil the rôle for which it was created, since it left unanswered all questions regarding the reliability of each perception.

Reid attacked the doctrine of ideas by appealing to common sense: the man in the street is convinced that what he perceives is the thing itself and not an idea nor an impression. Reid also believed that ideas in the philosophical sense do not exist but are fictitious creations that simply do not need to be postulated. Locke and Hume had based their notions on certain assumptions regarding elements of knowledge (for Locke, simple ideas and for Hume, impressions) and they had then gone on to regard knowledge as the result of combining these elementary data with the perception of their similarities and differences. However, Reid believed that the so-called elementary data were nothing more than the result of analysis.

To bring some clarity into the discussion, Reid made a distinction between *sensation* and *perception*. A *sensation* is, according to Reid, a feeling that exists exclusively in the mind of the perceiver.[1] Sensations have nothing in common with external objects, of which they are merely signs. By way of example of a sentence indicating a sensation Reid gave "I feel a pain"; the sentence "I see a tree" would, on the contrary, indicate a perception.[2] Reid held that a *perception* always involves an object distinct from the action by which it is perceived.[3] The perception of an object implies a concept of its form and a belief in its current existence. This belief is a consequence of our constitution and not of argumentation: "There is no reasoning in perception

[1] Reid [1764] 1970, ch. VI, §xx: "[a sensation] appears to be something which can have no existence but in a sentient mind, no distinction from the act of the mind by which it is felt."

[2] Reid [1764] 1970, ch. VI, §xx.

[3] Reid [1764] 1970, ch. VI, §xx: "Perception [...] hath always an object distinct from the act by which it is perceived."

[...] The belief which is implied in it, is the effect of instinct."[1] In contrast to Berkeley, Reid believed that an object can exist whether or not it be perceived.

Reid drew a distinction between two types of perception: some are *natural and original*, others are *acquired* through experience.[2] It is true of all senses to say that we have far more acquired than original perceptions, and it is especially true of sight:

> By this sense we perceive originally the visible figure and colour of bodies only, and their visible place; but we learn to perceive by the eye, almost every thing which we can perceive by touch. The original perceptions of this sense, serve only as signs to introduce the acquired.[3]

Like Locke and Berkeley, Reid drew a comparison between perception and *language*. Our original perceptions are, he said, analogous to natural languages, whereas acquired perceptions could be compared to artificial languages. The signs employed in original perceptions are observations; the signs used in natural languages are facial expressions, physical gestures and vocal modulations. These signs are universal and the capacity to interpret them is not acquired but inborn. In acquired perceptions the signs are either sensations or things that we perceive through means of sensations. In artificial languages the signs are articulated sounds. In both cases we discover the relationship through experience.[4]

Reid believed that perception must not only be distinguished from sensation but also from the *knowledge* of the objects of sense which we have acquired by reasoning. "When I look at the moon, I perceive her to be sometimes circular, sometimes horned, and sometimes gibbous. [...] from these various appearances of her enlightened part, I infer that she is really of a spherical figure. This conclusion is not obtained by simple perception, but by reasoning."[5]

Reid regarded sight as the noblest of our senses, but despite this he believed that only a fraction of the knowledge acquired by sight could not be communicated to a person blind from birth. To clarify this notion he drew a distinction between the visual appearance of things and the things which are suggested by that appearance:

---

[1] Reid [1764] 1970, ch. VI, §xx.

[2] Reid [1764] 1970, ch. VI, §xxiv: "The original perceptions which nature gave [children] are few, and insufficient for the purpose of life; and therefore she made them capable of acquiring many more perceptions by habit."

[3] Reid [1764] 1970, ch. VI, §xx.

[4] Reid [1764] 1970, ch. VI, §xxiv.

[5] Reid [1764] 1970, ch. VI, §xx.

> the visible appearance of objects is hardly ever regarded by us. It [...] serves only as a sign to introduce to the mind something else, which may be distinctly conceived by those who never saw.[1]

A book has various appearances, depending on the distance and the position from which it is observed. But *habit* has taught us that we should regard it as one and the same book: "overlooking the appearance, we immediately conceive the real figure, distance, and position of the body, of which its visible or perspective appearance is a sign and indication."[2] Reid thus agreed with Berkeley's "just and important observation" that the visual appearances of objects constitute a kind of language used by nature to inform us of distance, size and shape. Whenever we hear a familiar language we do not, after all, take notice of the sounds but of the meaning of the words, of the things indicated.

Reid believed that the visual form of an object is used only as a *sign* of the real figure. He put it this way: When I use my original powers of sight to look at a globe standing before me, I perceive only something of a circular form, variously coloured. The visible figure has no distance from the eye, no convexity, nor is it three-dimensional. But once I have *learnt* to perceive the distance of every part of this object from the eye, this perception gives it convexity and adds a third dimension.[3]

Reid attempted to reveal the characteristics of visual forms by introducing the *Idomenians*, creatures who (like Berkeley's *unbodied spirit*) would possess only the faculty of sight.[4]

> The being we have supposed having no conception of a third dimension, his visible figures have length and breadth indeed; but thickness is neither included nor excluded, being a thing of which he has no conception. And therefore visible figures, although they have length and breadth, as surfaces have, yet they are neither plain surfaces nor curve surfaces.[5]

The geometry developed by the Idomenians would, stated Reid, be non-Euclidean.[6]

---

[1] Reid [1764] 1970, ch. VI, §ii. The only profession for which it would be required to draw the distinction alluded to is, said Reid, that of a painter. See Reid [1764] 1970, ch. VI, §iii.

[2] Reid [1764] 1970, ch. VI, §ii.

[3] Reid [1764] 1970, ch. VI, §xxiii.

[4] The Idomenians can be regarded as the predecessors of the nineteenth-century Flatlanders invented by Edwin A. Abbott (Abbott 1884).

[5] Reid [1764] 1970, ch. VI, §ix.

[6] See also Daniels 1970.

Reid believed that while a blind person gaining his sight would perceive the same visible appearances of objects as we do, he would not understand their language; he would only take note of the signs without realising their significance. The blind man in Molyneux's question would therefore not know which of the two objects was the sphere and which the cube if he only used his *faculty of sight*:

> To a man newly made to see, the visible appearance of objects would be the same as to us; but he would see nothing at all of their real dimensions, as we do. He could form no conjecture, by means of his sight only, how many inches or feet they were in length, breadth or thickness. He could perceive little or nothing of their real figure; nor could he discern that this was a cube, that a sphere; that this was a cone, and that a cylinder. His eye could not inform him, that this object was near, and that more remote [...] In a word, his eyes, though ever so perfect, would at first give him almost no information of things without him.[1]

As we will see in the next section, Reid believed that the result would be different if the man was capable of *mathematical reasoning.*

Although there are other philosophers who gave a negative answer to Molyneux's question, the four already dealt with above will suffice here, since the philosophical analyses proposed by the rest do not differ essentially from those proposed by Molyneux, Locke, Berkeley and Reid. To the extent that these other philosophers based their conclusion (partly) on empirical data, they will be dealt with at a later stage.

Molyneux wrote to Locke that "upon Discourse with several concerning your Book and Notions, I have proposed [my Problem] to Diverse very Ingenious Men, and could hardly ever Meet with One that at first dash would give me the Answer to it, which I think true; till by hearing My Reasons they were Convinced."[2] What were the grounds advanced by these learned men for stating that a person blind from birth cured of blindness would be able to use no more than his faculty of sight to state which was the sphere and which the cube?

---

[1] Reid [1764] 1970, ch. VI, §iii.

[2] Molyneux to Locke, 2 March 1692/3, in Locke 1976–1990, vol. 4 (1979), no. 1609.

3 Positive Answers to Molyneux's Question

The first recorded example of a positive answer to Molyneux's question is that given by Edward Synge (1659–1741), later archbishop of Tuam.[1] Francis Quayle, prebendary of Brigown (Cloyne), laid out Molyneux's problem to Synge in the course of a visit to a friend who was a preacher. Synge became so obsessed by it that he could hardly put it out of his thoughts. On Friday 6 September 1695, the day after the visit, he wrote a letter to Quayle detailing his thoughts on the problem. Quayle sent the letter on to Molyneux who, in his turn, passed on a copy to Locke as an illustration of his statement that the problem had placed many scholars in an awkward position.[2]

In his letter Synge distinguished between various kinds of concepts: "I call every notion of any thing which a man entertains, an *Idea*, but that notion only, which a man entertains of a visible thing as it is visible I call an *image*."[3] According to Synge a person born blind can have a (tactile) *idea* of a sphere and a cube and of some difference that exists between them, but he cannot have any (visual) *image* of them. The idea he will have of the sphere is that of an object which is the same all over. The idea he forms of a cube is that of an object not the same on every surface, for in one place he will feel a smooth surface, in another the sharp point of an angle and in a third a long edge reaching from one angle to another.

If the man gains his powers of sight and is shown a sphere and a cube he will, according to Synge, immediately have different images of the sphere and the cube. The image he gets of the sphere is that of an object the same on all sides; the image of the cube will be that of an object not the same all over. The man will perceive *agreement* between the idea and the image of the sphere and *difference* between the idea of the sphere and the image of the cube (and *vice-versa*). These agreements and differences will, stated Synge, enable him to say which is the sphere and which the cube.[4]

---

[1] Edward Synge, bishop of Raphoe (1714) and archbishop of Tuam (1716), was particularly well known for his *Gentleman's Religion* (Synge 1693).

[2] Synge to Quayle, 6 September 1695, in Locke 1976–1990, vol. 5 (1979), no. 1984. Molyneux enclosed a copy of Synge's letter with the letter he wrote to Locke on Tuesday 24 December 1695. Synge's letter was published in Locke 1708.

[3] Synge to Quayle, 6 September 1695, in Locke 1976–1990, vol. 5 (1979), no. 1984.

[4] In fact this is what Synge wrote: "*if* immediately upon the sight of the globe and the cube there be ground enough for such a person Clearly to perceive the Agreement and the difference between his preconceived *ideas* and the newly conceived *images* of those figures *then* he may be able to know which is the globe and which is the Cube." [My italics *if* and *then*.]

In his letter, Molyneux said that Locke would very easily be able to uncover "by what false steps this Gentleman is lead into his Error."[1] Locke replied:

> I see by Mr. S's answer to that which was originally your question, how hard it is, even for ingenious men to free themselves from the anticipations of sense. The first step towards knowledge is to have clear and distinct ideas; which I have just reason every day more and more, to think few men ever have, or think themselves to want; which is one great cause of that infinite jargon and nonsense which so pesters the world.[2]

Locke's laudable attempts to acquire true knowledge on the basis of clear ideas were, however, unable to turn aside accusations that he himself had used erroneous language and incorrect reasoning. One of his critics was Henry Lee.

Twelve years after the publication of Locke's *Essay* Lee published his *Anti-Scepticism: or, Notes upon each Chapter of Mr. Lock's Essay concerning Humane Understanding* (1702). The work was first and foremost designed as an aid to Lee's two sons, who were studying philosophy. As the title would lead one to suspect, Lee accused Locke of scepticism: "The Idea of one thing being no more an evidence of its real Existence, than its real Existence is of a person's having the Idea of it.—And this is one reason why I conceive those Ideal Principles must involve us in an endless Scepticism."[3] Lee was of the opinion that simple ideas do not exist and that Locke's principles tended to undermine the foundations of natural and revealed religion.

Lee agreed with Locke that the senses are corrected by the judgment, but he doubted whether Locke's example of a uniformly coloured sphere demonstrated this. What Lee wondered was whether experience is required to recognise the difference between flat and spherical objects. He was convinced that the retinal images of a (flat) circle and a sphere are round. He also believed firmly that the image projected onto the retina by a flat surface is identical at all points, whereas the image of a spherical surface is not identical since shadows make a weaker impression than parts clearly illuminated. Lee believed that by nature we immediately perceive the illuminated parts of a sphere to be closer to us and the shaded parts to be further away.[4] He did not accept that experience was required to learn that shade signified shape, as Locke had suggested, "for the parts, from which the shade comes, will by the *nat-*

---

[1] Molyneux to Locke, 24 December 1695, in Locke 1976–1990, vol. 5 (1979), no. 1984.
[2] Locke to Molyneux, 5 April 1695, in Locke 1976–1990, vol. 5 (1979), no. 2059.
[3] Lee 1702, Preface.
[4] This only applies to front lighting.

ural Structure of the Eye appear farther off than otherwise they would, and so the whole Superficies of the Globe appear *protuberant*, as it really is."[1]

Regarding Molyneux's problem Lee wondered "whether there be not constituted in Nature a *necessary* Connexion between a certain *Motion* upon the organs of *Touch*, and a certain Perception, and a certain Figure at the bottom of the Eye, and that same Perception."[2] Lee thought that there must be a relationship of this nature, otherwise we would never be able to perceive one and the same object by both touch and sight. If the relationship did not exist, this would mean that nature had interposed a separation between the various senses, and Lee could not believe that this was the case. As we have seen, Berkeley proposed such a separation. He believed that there was no question of one single object being capable of being perceived by both touch and sight, but rather was it a question of two different objects. But even if there was a necessary relationship as proposed by Lee, the question would still remain as to whether a person born blind gaining his sight would immediately perceive the relationship.

Using a variant on Molyneux's problem Lee attempted to suggest that this was the case and that experience had no part to play. A person born blind who had learnt to distinguish between a saucer and a plate, which vary only in *size*, would—believed Lee—be able to see which was the larger and which the smaller without recourse to experience.[3] In the same way, the man would be able to see that in the case of a cube the distance from the centre point to points on the edge shows variations, whereas the radius of the sphere remains constant. Lee therefore believed that "the Author of Nature has annex'd the *same* Mode of Perception to a certain *Motion* upon the Organs of *Touch* as there is a certain *Figure* in the Eye."[4] The person born blind would only have to learn the *name* of the mode. In view of the fact that he had done that using the sense of touch, as Molyneux had assumed, Lee believed that he would be able to identify the sphere and the cube.

Locke's *Essay* met with resistance not only in England but also on the Continent. It persuaded the versatile scholar Gottfried Wilhelm Leibniz (1646–1716) to write a critical commentary, a document on the basis of which Leibniz attempted some time around 1695 to arrange discussions with Locke, but Locke did not react. In 1700 the French translation of Locke's *Essay* was published,[5] which persuaded Leibniz that he should lay out his criticism for a wider audience in an attractive literary form—as a dialogue. The result, the

---

[1] Lee 1702, bk. II, ch. ix, §8. This is incorrect.

[2] Lee 1702, bk. II, ch. ix, §8.

[3] This only applies when they are at the same distance from the perceiver.

[4] Lee 1702, bk. II, ch. ix, §8.

[5] Locke 1700.

Nouveaux essais sur l'entendement humain, was finished in 1704, but Locke's death in the same year took away any desire on the part of Leibniz to publish the work.[1] The first edition of the *Nouveaux essais* did not appear until 1765, almost fifty years after Leibniz himself had died.

It is no secret that Leibniz in no way shared Locke's empiricist ideas. He believed that in addition to knowledge which we acquire by experience there are also eternal and necessary truths with which we are born. Thus Leibniz changed Locke's statement saying that all our knowledge originates in experience into the saying: "*Nihil est in intellectu, quod non fuerit in sensu,* excipe: *nisi ipse intellectus.*"[2] Leibniz rejected the comparison Locke made between human understanding and a *tabula rasa* since this notion would lead to materialism. Instead, he compared human understanding to a block of marble, the veins of which represent virtual, inborn knowledge.

Leibniz agreed with Locke that unconscious judgments can occur in perception.[3] The system suggested by Leibniz was founded on the notion of *petites perceptions* or *perceptions insensibles*, tiny perceptions recorded unconsciously. He also distinguished conscious perceptions, known as *apperceptions*.

Like Lee's *Anti-Scepticism* the *Nouveaux essais* followed in the footsteps of Locke's *Essay.* At the appropriate place in the book, Philalèthe (Locke's spokesman) explained Molyneux's problem to Théophile. Théophile, defender of the opinions of Leibniz, replied:

> je crois que, supposé que l'aveugle sache que ces deux figures qu'il voit sont celles du cube et du globe, il pourra les discerner, et dire sans toucher. Ceci est le globe, ceci le cube.[4]

In reply to Philalèthe's remark that he had given the wrong solution, Théophile answered that his view did not differ as much from that of Molyneux or Locke as might seem to be the case. He pointed out that within the question there was a hidden condition that the blind person should only be able to make

---

[1] In 1706 Leibniz wrote to Thomas Burnet: "La mort de M. Locke m'a ôté l'envie de publier mes remarques sur ces ouvrages; j'aime mieux publier maintenant mes pensées indépendamment de celles d'un autre." Quoted by J. Brunschwig in his introduction to Leibniz [1765] 1966, p. 15.

[2] Leibniz [1765] 1966, bk. II, ch. i, §2.

[3] Leibniz accused Locke of failing to adhere to his own programme: "Il semble que notre habile auteur prétend qu'il n'y a rien de virtuel en nous et même rien dont nous ne nous apercevions toujours actuellement; mais il ne peut pas le prendre à la rigueur, autrement son sentiment serait trop paradoxe, puisque encore les habitudes acquises et les provisions de notre mémoire ne sont pas toujours aperçues [...] Il limite aussi sa thèse en d'autres endroits, en disant qu'il n'y a rien en nous dont nous ne nous soyons au moins aperçus autrefois." Leibniz [1765] 1966, *Préface.*

[4] Leibniz [1765] 1966, bk. II, ch. ix, §8.

a *distinction* between the two objects. For the man would know beforehand that he was to be shown a sphere and a cube. According to Théophile it was therefore without question that he would be able to distinguish them "[par] les principes de la raison, joints à ce que l'attouchement lui a fourni auparavant de connaissance sensuelle."[1] The reason for this conviction was that a sphere has no points which are distinct from other points because its surface is flat and without angles, while a cube has eight points which are distinct from each other.

Leibniz thus placed arguments in Théophile's mouth similar to those advanced by Synge. He subsequently had Théophile state that if this means of distinguishing forms were not to exist, a blind person would not be capable of understanding the basics of geometry by means of touch, whereas we know that persons blind from birth can certainly do this. In contrast to Berkeley Leibniz believed that a paralytic could, indeed, do geometry. He even stated that geometry is mostly learnt by sight alone. And if the geometry of a blind person were to be compared with that of a paralytic, we would see that both have arrived at the same *ideas*, though they would not have any common *images*:

> Et il faut que ces deux géométries, celle de l'aveugle et celle du paralytique, se rencontrent et s'accordent et même reviennent aux mêmes idées, quoiqu'il n'y ait point d'images communes.[2]

What Leibniz says here is of major importance. He makes an explicit distinction between images and ideas: "Ce qui fait encore voir combien il faut distinguer *les images des idées exactes*, qui consistent dans les définitions."[3] According to Leibniz a definition can be given of ideas—such as ideas of shape—which are acquired by means of more than one sense:

> Les idées qu'on dit venir de plus d'un sens, comme celle de l'espace, figure, mouvement, repos, sont plutôt du sens commun, c'est-à-dire de l'esprit-même, car ce sont des idées de l'entendement pur, mais qui ont du rapport à l'extérieur, et que les sens font apercevoir; aussi sont-elles capables de définitions et de démonstrations.[4]

When it is possible to give a definition of the form of an object, Leibniz believed, it is then that we have an *exact idea* of the object. If, however, the nature and characteristics of an object are not known, we can have only an *image* or an *unclear idea* (*idée confuse*) of it. Thus a mathematician can have

[1] Leibniz [1765] 1966, bk. II, ch. ix, §8.
[2] Leibniz [1765] 1966, bk. II, ch. ix, §8.
[3] Leibniz [1765] 1966, bk. II, ch. ix, §8.
[4] Leibniz [1765] 1966, bk. II, ch. v.

an exact idea of a nine-sided or ten-sided figure without being immediately capable of distinguishing them one from the other by sight. A workman or an engineer, by way of contrast, can possess a very clear image of a ten-sided figure without having an exact idea of it.[1]

A person blind from birth acquiring the faculty of sight would, said Leibniz, be capable of distinguishing a sphere from a cube using his understanding and the knowledge acquired through the sense of touch, or with the aid of exact ideas of the forms of both objects. But if the man were *not* to be told beforehand that he would be shown a sphere and a cube, Leibniz believed that he would not have any idea that he was seeing three-dimensional objects:

> je réponds qu'il les discernera, comme je viens de dire, si quelqu'un l'avertit que l'une ou l'autre des apparences ou perceptions qu'il en aura appartient au cube et au globe; mais sans cette instruction préalable, j'avoue qu'il ne s'avisera pas d'abord de penser que ces espèces de peintures qu'il s'en fera dans le fond de ses yeux, et qui pourraient venir d'une plate peinture sur la table, représentent des corps.[2]

This opinion is, indeed, very close to what Locke had stated.

Leibniz emphatically added, through the mouth of Théophile, that he was not speaking of what the man would immediately do when amazed and confused by the novelty, or when he was scarcely used to drawing consequences: "je ne parle pas de ce qu'il fera peut-être en effet et sur-le-champ, étant ébloui et confondu par la nouveauté ou d'ailleurs peu accoutumé à tirer des conséquences."[3] In this, Leibniz was of the same opinion as Berkeley.

Francis Hutcheson (1694–1747), an Irish philosopher known particularly for his theory of *moral sense* in ethics, gave an answer to Molyneux's question which is remarkably similar to that given by Leibniz. On 6 September 1727 Hutcheson wrote a letter to William Mace of Gresham College in which he dealt with Berkeley's immaterialism and, partly in response to his point of view, gave a positive answer to Molyneux's question. The letter was first published in 1788.[4]

As already mentioned, Berkeley had stated in his *Essay* that either visual extension and forms differ from tactile extension and forms or that the solution proposed by Molyneux and Locke was incorrect.[5] Berkeley was of the opinion that the former was the case; Hutcheson held to the latter: "Messrs. Locke and Molyneux are both wrong about the cube and sphere proposed to a blind

[1] Leibniz [1765] 1966, bk. II, ch. xxix, §13.
[2] Leibniz [1765] 1966, bk. II, ch. ix, §8.
[3] Leibniz [1765] 1966, bk. II, ch. ix, §8.
[4] Hutcheson [1727] 1788.
[5] Berkeley [1709] 1975, §133.

man restored to sight. He would not at first know the sphere from a shaded plain surface by a view from above; but a side view would discover the equal uniform relievo in one, and the cubic one in the other."[1]

According to Hutcheson we can judge by touch, with eyes closed, what the visual extension of a touched object will be when we open our eyes, but we cannot determine by touch which colour we will see:

> which shews visible and tangible extension to be really the same idea, or to have one idea common, viz. the extension; though the purely tangible and visible perceptions are quite disparate.[2]

This statement clearly recalls the statement made by Leibniz that visual and tactile images of forms differ from one another, although their exact ideas are the same. Both philosophers meant that while visual and tactile sensations of forms are different, the accompanying concepts are the same or have something in common.[3]

Like Leibniz, Hutcheson believed that a blind person would understand nothing of geometry if visual and tactile ideas of extension were totally different, as Berkeley had suggested. However, blind people do in fact understand geometry—using, for example, wooden shapes—and thus visual and tactile ideas must, he believed, be the same.

Hutcheson wanted to illustrate the erroneous nature of Berkeley's thesis using a thought experiment inspired by Molyneux's problem. Imagine a person, proposed Hutcheson, both paralytic and blind, with no notion of either type of extension, yet having a keen sense of smell. Imagine further that an object exists which changes its smell with every change of its shape. The blind paralytic will give the various smells various names. Another person, this one sighted, will use the same names for the various shapes. This latter person will reason concerning the shapes or will formulate one of Euclid's theorems concerning the relationship between the sides. Would it be possible, asked Hutcheson, for the blind paralytic to agree with this? Would he

---

[1] Hutcheson [1727] 1788, p. 159. See also Berman 1974a. It is unclear why a view from above would lead to recognition while a side view would not.

[2] Hutcheson [1727] 1788, p. 159.

[3] Hutcheson stated that a sensation is accompanied by an idea; here he was speaking of "concomitant ideas." Hutcheson [1727] 1788, p. 158. See also Hutcheson 1728, p. 3, note: "Some *Ideas* are found accompanying the most different Sensations, which are not yet to be perceived separately from some *sensible Quality*; such are *Extension*, *Figure*, *Motion*, and *Rest*, which accompany the *Ideas* of Sight, or Colours, and yet may be perceived without them, as in the *Ideas* of *Touch*, at least if we move our Organs along the Parts of the Body touched. *Extension*, *Figure*, *Motion*, or *Rest* seem therefore to be more properly called *Ideas accompanying* the Sensations of *Sight* and *Touch*, than the Sensations of either of these Senses."

recognise the meaning by means of the smells? Hutcheson did not answer this rhetorical question. Possibly he was hoping that all right-thinking people would answer "no," since smells and visual forms and extension have nothing in common. But a blind person would certainly be able to speak about geometry with someone possessing only the sense of sight. This, stated Hutcheson, demonstrates that visual and tactile extension (and forms) possess an idea in common. And apparently he believed that a person born blind would immediately notice this idea when he first saw a sphere and a cube.

In the second edition of his *Essai philosophique sur l'âme des bêtes*, published in 1737, the Dutch preacher and philosopher David Renaud Boullier (1694–1759) proposed a solution to Molyneux's problem which was in line with those put forward by Leibniz and Hutcheson.[1] The aim of his *Essai* was to prove that animals possess a soul and to investigate the nature of this soul. Boullier's discussion of the perception of animals tempted him into enlarging on the nature of our sensations, which brought him to Molyneux's problem, which he called "Paradoxe de mr. Locke."

The answer given by Locke was, said Boullier, incorrect and "peu digne d'un Esprit aussi solide que le sien."[2] The reason why Locke's reasoning did not add up was, according to Boullier, as follows:

> L'idée du Globe vu, par exemple, et l'idée du Globe touché sont la même idée essentielle, quoi que diversement modifiée par différens accompagnemens de perceptions accessoires. Que mes mains touchent, ou que mes yeux voyent un Corps, l'idée de son étendue s'imprime également par ces deux voyes dans mon Esprit.[3]

Boullier therefore concluded that a person born blind gaining his sense of sight would discover the same difference between the sphere and the cube as he had already learnt through his sense of touch:

> je conclus que l'Aveugle guéri retrouvera bien-tôt, par le secours des yeux, la même différence entre le Globe et le Cube, qu'il ne connoissoit jusqu'alors que par l'attouchement.[4]

Despite Boullier's criticism of Locke, various aspects of their ideas seem to agree. Boullier, for instance, admitted that the faculty of sight is not perfect

[1] The first edition of this work appeared in 1728. I have not seen this edition.

[2] Boullier [1728] 1737, part II, ch. vi, §18.

[3] Boullier [1728] 1737, part II, ch. vi, §18.

[4] Boullier [1728] 1737, part II, ch. vi, §18. Here Boullier added that "la vue de ces deux Solides perfectionnera, rendra plus nette et plus vive l'idée qu'il en avoit déja reçue par un autre sens, et lui fera faire par conséquent un discernement plus exact et plus juste entre l'un et l'autre."

and that it has, like other senses, its illusions which other senses, experience and understanding have to correct. Sight, for example, would show us solid objects as flat and deformed:

> Selon la distance et l'aspect, les Corps sont vus sous une figure différente de la véritable. Les angles d'un Cube sont émoussez, certains côtez retrécis, un Globe nous paroît une surface platte et circulaire etc.[1]

However, Boullier believed that this does not prevent our sight from providing knowledge of the same properties as does our sense of touch and that it assures us of the identity of the object that we have perceived through these two senses.

Boullier was of the opinion that the example of the sphere and the cube was deceptive because sight does not give us immediately an idea of solidity, nor of the hardness or resistance of a body. He was the first to suggest taking a *square* and a *circle.* The blind person in question will acquire the idea of these two shapes and their properties viâ the sense of touch, said Boullier, and since the faculty of sight will reveal the same properties the man will be able to say which is the square and which the circle:

> La vue lui offre les mêmes rapports d'égalité entre les rayons du Cercle, la même uniformité de Courbure, les mêmes différences entre le Cercle et le Quarré, que le tact lui avoit déja fait apercevoir. Là-dessus il dira sans héziter, c'est un Cercle; et de l'autre figure, c'est un Quarré.[2]

Is it not asking too much to ascribe this type of reasoning to a blind person? Would we not be raising him to the level of a deep-thinking philosopher? No, affirmed Boullier, for I am merely proposing that he is capable of thinking, which in no way deviates from the terms in which the problem is laid out; moreover it is well known that blind people are good practitioners of geometry. What we should really be asking, according to Boullier, is whether the man would be capable of distinguishing the two objects from each other without first thinking deeply about it:

> C'est ce qui doit infailliblement arriver, ce me semble. Cette action de notre Âme qui réunit au même Objet les diverses Sensations qu'elle en reçoit se servant de leur concours pour se former une idée plus nette de cet objet est un raisonnement prompt, subtil, imperceptible, mais

---

[1] Boullier [1728] 1737, part II, ch. vi, §18.
[2] Boullier [1728] 1737, part II, ch. vi, §19.

> naturel, qui du Corps même que nous animons s'étend par analogie sur les autres Corps qui l'environnent.[1]

In this regard Boullier differed in his opinion from Leibniz.

The above shows that Molyneux's problem was subject to a variety of interpretations. There were, for instance, differences of opinion regarding the question of whether the person born blind would have to respond immediately or whether he would get the time to think about the problem. The London doctor James Jurin (1684–1750) explicitly brought up this and other questions.[2]

Jurin believed that Locke had understood Molyneux's question differently from the way Molyneux had meant it and Locke's judgment would therefore be totally different from that of Molyneux. Molyneux had, in fact, only forbidden the man to touch the objects. But he would have left him free to look at them time and again, to observe them from different angles and to use his memory and understanding.[3] Did Locke allow this? No, replied Jurin: "As soon as he is made to see, [Locke] requires him to pronounce, *with certainty at first sight*, which is the globe, which the cube; without giving him leave to take a second view, much less to recollect himself and to reason upon what he sees."[4]

The question is whether Locke's *at first sight* should be understood as strictly as Jurin believed. Other authors used the expression, while it is clear that they only meant to say "when the man has gained his faculty of sight." Jurin's remark is nonetheless interesting in that it could make a difference whether the man would be required to give an immediate answer or whether he would be given the time to reflect, and whether he would have to stand still or would be allowed to move his head or even his whole body.

Like Leibniz, Jurin believed that it would be totally impossible for the man *of himself* to distinguish a sphere from a cube by sight.[5] But if he *were to be told* that he would be given these objects to look at (and, said Jurin, this condition was part of Molyneux's question) then he would be able to do it.[6]

---

[1] Boullier [1728] 1737, part II, ch. vi, §19.

[2] Jurin was a well-known polemicist of his time. He had been engaged in a controversy with Berkeley: see Berkeley 1734, Jurin 1734, Berkeley 1735, Jurin 1735, and Breidert 1989.

[3] Jurin 1738, p. 28, §163.

[4] Jurin 1738, p. 28, §164.

[5] Jurin 1738, p. 28, §166: "That he shall be told, the two bodies he sees are one a globe and the other a cube; without which information it is no purpose to ask him which is the globe and which the cube."

[6] Jurin stated that the question would also imply that "the blind man shall by sight perceive the globe as one thing, distinct from the cube and all other bodies; and shall likewise perceive the cube as one thing distinct from the globe and all other bodies." Jurin 1738, p. 28, §166. However, it was precisely the question as to whether this would be the case.

If the man were to look at the two objects from various angles, he would note that in the one case he was receiving different sensations. In the other case he would always receive a similar sensation regardless of the side from which he looked at the object. If he were then to recall how a sphere and a cube felt, he would reason his way to the conclusion that a sphere is an object which is the same over its entire surface whereas a cube is not. Since our senses have not been given to us in order to deceive us, said Jurin, and different sensations are caused by different objects, the man would be able to conclude which was the sphere and which the cube. We have seen this sort of reasoning before, as proposed by Synge and others.[1] Finally, Jurin wrote that he had heard from Smith that the greatest blind philosopher ever to exist, Doctor Saunderson who—remarkably enough—taught optics, agreed with him.[2]

Reid agreed with the opinion put forward by Leibniz and Jurin that the person born blind would be able to identify the sphere and the cube if he were capable of (mathematical) reasoning. Reid had praised Berkeley for having seen that visual forms are signs of tactile forms. But he believed that Berkeley had pushed his thesis too far by assuming that there is no relationship at all between the size, the form and the position which we see and which we feel. Reid believed that there truly is a similarity, even a necessary, mathematical relationship between the visual form and size of an object and its actual form and size. In the case of a sighted person, the visual form leads by experience directly to the concept of its true form of which it is a sign. The true form and position can, according to Reid, be derived through mathematical reasoning from the visual form and the distance of the various components. In the same way, namely by mathematical reasoning, the visual form of an object can be derived from its real form, distance and position. Reid believed that this was possible not only for a sighted person but also for one blind from birth.[3]

In order to investigate whether he was right or not, Reid conducted a thought experiment: "let us suppose such a blind man as Dr. Saunderson, having all the knowledge and abilities which a blind man may have, suddenly made to see perfectly."[4] Then imagine, continued Reid, that the man is not given the opportunity to associate the ideas of sight with those of touch until such time as the ideas of sight begin to be somewhat familiar to him. When

---

[1] Joseph Priestley (1733–1804), the discoverer of oxygen, agreed with Jurin's answer to Molyneux's question: see Priestley 1772, pp. 720–725.

[2] Jurin 1738, p. 29, §170. Nicholas Saunderson (1682–1739) was one year old when he lost his sight as result of smallpox. He was Lucasian professor of mathematics and the subjects he taught included Newtonian philosophy, hydrostatics, astronomy, acoustics and optics. He practised palpable arithmetic using a calculating table which he had himself invented. Saunderson detailed his arithmetic in Saunderson 1740.

[3] Reid [1764] 1970, ch. VI, §vii.

[4] Reid [1764] 1970, ch. VI, §xi.

the first surprise caused by new objects has subsided, the man should be given time to study them in his mind and to compare them with the ideas obtained by touch; and especially to compare in his mind the visual form and extension with the form and extension in length and breadth with which he is already familiar thanks to the sense of touch. Reid believed that the man would be capable of perceiving that both have length and breadth: "he will perceive, that there may be visible as well as tangible circles, triangles, quadrilateral and multilateral figures."[1] This would be particularly true of small objects which can be perceived at a glance. In the case of larger objects, Reid believed, the properties of visual forms would differ from those of the flat surfaces they represent. Reid concluded that

> if Dr. Saunderson had been made to see, and attentively had viewed the figures of the first book of Euclid, he might, by thought and consideration, without touching them, have found out that they were the very figures he was before so well acquainted with by touch.[2]

If flat surfaces were to be looked at from an angle, this would be more difficult since the visual form then differs more from the tactile form. And, said Reid, the visual representation of three-dimensional forms would be less perfect because visual extension does not have three but only two dimensions. Nonetheless, we can say that there is some similarity, so Reid believed. And therefore Berkeley had committed a major error by assuming that there is absolutely no agreement between the extension, form and position which we see and which we perceive by touch.[3]

As far as the positive replies to Molyneux's question are concerned, we will limit our investigations to the above philosophers. Here too the same applies: others thinkers replied in similar ways. Those who supported their point of view with empirical data will be dealt with in a subsequent chapter.

4 Conclusions

All the philosophers we have so far examined made the basic assumption that tactile and visual *sensations* differ from one another, and that is an acceptable point of view. A number of them based their solution to Molyneux's problem on this point. Some philosophers were of the opinion that there is a *necessary relationship* between visual and tactile sensations of the form of objects. Those

[1] Reid [1764] 1970, ch. VI, §xi.
[2] Reid [1764] 1970, ch. VI, §xi.
[3] Reid [1764] 1970, ch. VI, §xi.

who, like Molyneux, believed that this relationship can only be found out through experience expected that the person born blind would be incapable of distinguishing a sphere from a cube by sight. However, those who, like Lee, were convinced that this relationship could be perceived immediately, gave positive answers to Molyneux's question. Those who adopted a third position, including Jurin and Reid, were of the opinion that the relationship could be deduced by (mathematical) reasoning; they too gave a positive answer. Berkeley and his disciples, however, believed that experience merely created an *arbitrary relationship* between visual and tactile perceptions. A person born blind would, they said, be unable to identify the sphere and the cube by sight.

Molyneux's problem was also solved on the basis of other considerations, related to the relationship between visual and tactile *concepts* of the form of objects. Some philosophers assumed that these concepts *differ* from one another and can be made to relate to one another either by experience or by understanding. Reasoning of this kind produced a negative and a positive response respectively.

Other scholars, however, believed that visual and tactile concepts of form are *essentially identical* or, at any rate, possess a *common* concept. Some of these people, including Boullier and Hutcheson, believed that this could be immediately perceived. Others, such as Leibniz, were of the opinion that this identity would only be brought to light by understanding. These philosophers were of the opinion that the person born blind restored to sight would be able to state (immediately or not) which of the objects was the sphere and which the cube.

In brief, there was a range of opinions which served as arguments in the solutions proposed to Molyneux's question. There was no unanimity as to what the correct solution could be. The most that can be said is that thinkers who were inclined towards rationalism tended to give a positive answer while the empiricists usually answered Molyneux's question in the negative.

The varying *interpretations* given to Molyneux's question hampered any attempts to reach a unanimous answer. Some philosophers assumed that the person born blind had to answer immediately while others thought it normal that he be allowed to use his understanding and to move around the objects. Some thought it would even make a difference if the man were to know that he was to be presented with a sphere and a cube. And, of course, the various options led to different answers.

Molyneux's problem inspired philosophers to invent further thought experiments designed to provide us with information about our perception. Berkeley, for example, introduced the notion of an *unbodied spirit*, while Hutcheson imagined a world in which the re-shaping of objects would be accompanied by different smells. Leibniz, moreover, pointed to the importance of studying the ideas of people lacking one or other of the senses, a theme which we will encounter again.

Up till this stage no-one had posed the question as to whether a person born blind could really be given the power of sight. Everyone regarded Molyneux's problem as a hypothetical question amenable to a solution by reasoning. It was not until it became known that certain people born blind could be given their sight by a cataract operation that the question arose as to whether the eyes could provide useful information immediately following the operation. From that moment attempts were made to test the solutions grounded in theory against experimental data.

In fact Berkeley was the first to perceive the importance of experience as the touchstone for his solution to Molyneux's problem. In the Appendix to the second edition of his *Essay* of 1710 he announced that shortly after the publication of the first edition of the *Essay* he had learnt that a man born blind in the vicinity of London had recovered his sight at the age of twenty:

> Such a one may be supposed a proper judge to decide how far some tenets laid down in several places of the foregoing essay are agreeable to truth, and if any curious person hath the opportunity of making proper interrogatories to him thereon, I should gladly see my notions either amended or confirmed by experience.[1]

The operation in question had been performed by a certain Roger Grant, an eye doctor who was regarded as a fraud and whose story was accorded no credibility.[2] Berkeley never referred to Grant's operation again; but he gave some attention to the report of a successful cataract operation published by William Cheselden in 1728.

As will be seen in the next chapter, after 1728 Molyneux's problem was usually mentioned in the same breath as the Cheselden report. It constituted a source of arguments to support various solutions proposed to Molyneux's problem. Because it appeared difficult to interpret the report unambiguously, it also came to represent a stimulus to formulate criteria to which an adequate experimental test of Molyneux's problem would have to conform.

---

[1] Berkeley [1710] 1975, p. 59.

[2] A description of this eye operation can be found in *The Tatler*; see [Steele] [1709] 1898. An anonymous pamphlet allegedly written by Roger Grant (Grant [pseud.] 1709) states that Grant (?–1724) was a Baptist preacher, had been a cobbler and was illiterate. No. 444 of the *Spectator* (30 July 1712) satirically describes Grant as "putting out eyes with great success." See [Steele] [1709] 1898, pp. 42–43, note by the editor, and Pastore 1971, pp. 388–389, note 24.

Chapter Four

THE FIRST EXPERIMENTAL DATA

1 Cheselden's Operation

IN the *Philosophical Transactions* of the Royal Society of 1728 an article was published that was to play an important part in the discussion surrounding Molyneux's problem. The author of the article was the famous English surgeon and anatomist William Cheselden and was entitled "An Account of some Observations made by a young Gentleman, who was born blind, or lost his Sight so early, that he had no Remembrance of ever having seen, and was couch'd between 13 and 14 Years of Age."[1]

As the title indicates, Cheselden had performed a successful operation on a young man who, from birth or at least from a very young age, had suffered from cataract. For this Cheselden had performed two cataract operations: first he operated on one of the boy's eyes and then, a year later, on the other.[2] In his report Cheselden described the condition of the patient's eyes before the operation and he also gave a detailed description of what the boy could and could not see following surgery and how he reacted to his new visual impressions.

As could have been anticipated, those interested in Molyneux's problem were extremely eager to find out the results of Cheselden's operation, since it promised to provide them with arguments that would settle the dispute raging between the various parties. As far as we can ascertain, Cheselden himself did not know of Molyneux's problem. He referred to neither Molyneux nor Locke (nor to other parties to the discussion) and he did not show his patient a sphere or a cube.

[1] William Cheselden (1688–1752) was famous not only for the cataract operation alluded to here but also for his "lateral operation for the stone" (an operation to remove kidney stone) (1727) which was attended by surgeons from the whole of Europe. Cheselden was the discoverer of the artificial pupil. He wrote two seminal works: *The Anatomy of the Human Body* (1713) and *Osteographia, or Anatomy of the Bones* (1733). The *Anatomy* is full of magnificent copperplate engravings produced using the camera obscura. See Cope 1953 and Hausmann 1989.

[2] The following section explains what a cataract operation is.

Cheselden did not write his report until after the second operation, probably basing his writing on what he could remember and not on notes made at the time. In general he set down the observations made by his patient without further comment. Only a few times did he report *verbatim* what the boy had said. One of these passages caused great confusion for decades and was the occasion of a great deal of criticism.

Because Cheselden's report is of major significance, I quote it in full. Then I take the opportunity to describe cataract as it actually is and the sorts of cataract operation usual in the eighteenth century. Following that I shall seek out the information contained in Cheselden's report that can be related to Molyneux's problem. But first let us hear the voice of William Cheselden, "F. R. S. Surgeon to her *Majesty*, and to St. *Thomas*'s Hospital":

> Tho' we say of the Gentleman that he was blind, as we do of all People who have Ripe Cataracts, yet they are never so blind from that Cause, but that they can discern Day from Night; and for the most Part in a strong Light, distinguish Black, White, and Scarlet; but they cannot perceive the Shape of any thing; for the Light by which these Perceptions are made, being let in obliquely thro' the aqueous Humour, or the anterior Surface of the Chrystalline (by which the Rays cannot be brought into a *Focus* upon the *Retina*) they can discern in no other Manner, than a sound Eye can thro' a Glass of broken Jelly, where a great Variety of Surfaces so differently refract the Light, that the several distinct Pencils of Rays cannot be collected by the Eye into their proper *Foci*; wherefore the Shape of an Object in such a Case, cannot be at all discern'd, tho' the Colour may: And thus it was with this young Gentleman, who though he knew these Colours asunder in a good Light; yet when he saw them after he was couch'd, the faint Ideas he had of them before, were not sufficient for him to know them by afterwards; and therefore he did not think them the same, which he had known before by those Names. Now Scarlet he thought the most beautiful of all Colours, and of others the most gay were the most pleasing, whereas the first Time he saw Black, it gave him great Uneasiness, yet after a little Time he was reconcil'd to it; but some Months after, seeing by Accident a Negroe Woman, he was struck with great Horror at the Sight.
>
> When he first saw, he was so far from making any Judgment about Distances, that he thought all Objects whatever touch'd his Eyes, (as he express'd it) as what he felt, did his Skin; and thought no Objects so agreeable as those which were smooth and regular, tho' he could form no Judgment of their Shape, or guess what it was in any Object that was pleasing to him: He knew not the Shape of any

Thing, nor any one Thing from another, however different in Shape, or Magnitude; but upon being told what Things were, whose Form he before knew from feeling, he would carefully observe, that he might know them again; but having too many Objects to learn at once, he forgot many of them; and (as he said) at first he learn'd to know, and again forgot a thousand Things in a Day. One Particular only (tho' it may appear trifling) I will relate; Having often forgot which was the Cat, and which the Dog, he was asham'd to ask; but catching the Cat (which he knew by feeling) he was observ'd to look at her stedfastly, and then setting her down, said, So Puss! I shall know you another Time. He was very much surpriz'd, that those Things which he had lik'd best, did not appear most agreeable to his Eyes, expecting those Persons would appear most beautiful that he lov'd most, and such Things to be most agreeable to his Sight that were so to his Taste. We thought he soon knew what Pictures represented, which were shew'd to him, but we found afterwards we were mistaken; for about two Months after he was couch'd, he discovered at once, they represented solid Bodies; when to that Time he consider'd them only as Party-colour'd Planes, or Surfaces diversified with a Variety of Paint; but even then he was no less surpriz'd, expecting the Pictures would feel like the Things they represented, and was amaz'd when he found those Parts, which by their Light and Shadow appear'd now round and uneven, felt only flat like the rest; and ask'd which was the lying Sense, Feeling or Seeing?

Being shewn his Father's Picture in a Locket at his Mother's watch, and told what it was, he acknowledged a Likeness, but was vastly surpriz'd; asking, how it could be, that a large Face could be express'd in so little Room, saying, It should have seem'd as impossible to him, as to put a Bushel of any thing into a Pint.

At first, he could bear but very little Sight, and the Things he saw, he thought extreamly large; but upon seeing Things larger, those first seen he conceiv'd less, never being able to imagine any Lines beyond the Bounds he saw; the Room he was in he said, he knew to be but Part of the House, yet he could not conceive that the whole House could look bigger. Before he was couch'd, he expected little Advantage from Seeing, worth undergoing an Operation for, except reading and writing; for he said, He thought he could have no more Pleasure in walking abroad than he had in the Garden, which he could do safely and readily. And even Blindness he observ'd, had this Advantage, that he could go any where in the Dark much better than those who can see; and after he had seen, he did not soon lose this Quality, nor desire a Light to go about the House in the Night. He said, every new Object was a new Delight, and the Pleasure was so great, that he wanted Ways to express it; but his Gratitude to his Operator he could not conceal, never seeing him for some Time without Tears of Joy in his Eyes, and other Marks of Affection: And if he did not happen to come at any Time when he was expected, he would be so griev'd, that he could not forbear crying at his Disappointment. A Year after first Seeing, being carried upon *Epsom Downs*, and observing a large Prospect, he was exceedingly delighted with it, and call'd it a new Kind of Seeing. And now being lately couch'd of his other Eye, he says, that Objects at first appear'd large to this Eye, but not so large as they did at first to the other, and looking upon the same Object with both Eyes, he thought it look'd about twice as large as with the first couch'd Eye only, but not Double, that we can any Ways discover.[1]

Cheselden's report can be regarded as one of the first and most important of a series of similar reports.[2] It became well known because it was often referred to. Pastore called it "the most celebrated case study in the history of science until the early case studies of Freud came along at the beginning of the twentieth century."[3] Its great influence was largely due to the fact that it was sufficiently detailed to stand comparison with later reports over a period of several decades. In most of these cases the first visual tests were not sufficiently well prepared, so that even when the post-operative conditions were favourable the first moment of seeing was able to pass unremarked.

From 1728 Molyneux's problem was usually coupled to Cheselden's report. As we shall see, those who already believed on theoretical grounds that a person born blind would be unable to distinguish a sphere from a cube regarded the report as confirmation of their ideas, while those holding the opposite opinion usually issued commentaries on it. Before becoming embroiled in this controversy, however, let us examine the exact nature of cataract and see what kind of cataract operations were being performed in the eighteenth century.

[1] Cheselden 1728, pp. 447–450. Cheselden included his report in later editions of *The Anatomy* and added the following sentence: "I have couched several others who were born blind, whose observations were of the same kind; but they being younger, none of them gave so full an account as this gentleman." Quoted by Morgan 1977, p. 21.

[2] Von Senden [1932] 1960, pp. 326–335, supplies an extensive list of reported cataract operations, beginning with that carried out by Ammar (c. 1000, Egypt) up to this century.

[3] Pastore 1971, p. 99.

2 Cataract

In order better to understand what cataract is and how it can be cured, we must first examine a healthy eye. As is well known, the eyeball consists of an anterior chamber, a posterior chamber, the lens and the vitreous body. The chambers of the eye are separated by the iris. The lens is convex on both sides and thanks to its transparency and capacity for changing shape is an important component of the eye's optical system. The whole is encased in a triple layer of tissue. The outer layer (tunica fibrosa) consists of the cornea—at the front—and the sclera, which covers the rest of the eyeball. The cornea consists of clear, transparent tissue that makes the greatest contribution to the total refraction of light in the eye. (The refraction caused by the lens is much less than that caused by the cornea.) The second layer (tunica vasculosa) consists of the iris at the front and the chorioidea at the back of the eye, plus the corpus ciliare which binds the two together. At the front of the eye the iris constitutes the pupil. The innermost layer covering the eye is the retina.

Until some time in the seventeenth century cataract was regarded as an eye disease caused by a membrane which grew as a result of a thickening of the watery anterior chamber (between cornea and iris) and which was situated between the pupil and the lens.[1] In the sixth decade of the seventeenth century, however, the French eye specialist Lasnier discovered that cataract is a disturbance of the *lens*.[2] His discovery was confirmed in the early eighteenth century by Maître-Jan and Brisseau.[3] Brisseau made the following remark about Lasnier:

> Nous ne sommes pourtant pas les premiers qui en ont parlé; & l'on a sceu depuis que M. Lasnier habile Chirurgien de Paris & Oculiste, avoit fait la Même découverte il y a plus de 40 ans [...] Mais nous n'en sommes pas moins les inventeurs, puisqu'elle estoit absolument tombée dans l'oubli; que de nostre temps on n'en fait aucune mention dans les Traités, ni dans les Cours d'anatomie & d'operations; & que l'Académie même l'a regardée comme une nouveauté.[4]

---

[1] Münchow 1984, pp. 262–283.

[2] It is unclear when exactly Lasnier made his discovery. [Trévoux] 1714 contains the name of the anatomist Remigius L'Asnier, who died in 1690. This work also states that L'Asnier did not regard cataract as a membrane between the cornea and lens but as a deviation in the lens itself. See Münchow 1984, p. 270.

[3] Antoine Maître-Jan announced his discovery in Maître-Jan 1707. The *Académie Royale des Sciences* had, however, already given its imprimatur on 11 April 1704. Michel Brisseau's discovery was made known to the *Académie* on 17 November 1705. By way of a rare exception in the history of medicine, the two doctors did not engage in conflicting claims regarding the discovery. See Münchow 1984, p. 263.

[4] Brisseau 1709, as quoted in Münchow 1984, p. 264.

The fact that this *nouveauté* was not immediately granted the recognition it deserved can be seen in the numerous objections to it in the *Mémoires de l'Académie Royale des Sciences.*

3 Cataract Operations

In a healthy eye (with a transparent lens) the image of an object in the field of vision is focused on the retina by means of accommodation. But when the lens is opaque this is no longer possible.[1] Cataract can be congenital or can come into being at a later age. It can also occur in various degrees, depending on the extent of the opaqueness. Like Cheselden's patient, sufferers can usually perceive light but not shape.

There are various ways of removing an opaque lens so that it no longer blocks off the light. The oldest and simplest method is *reclination* of the cataract, known as 'couching' or 'depressing' the cataract. It involves the eye surgeon using a needle to move aside the opaque lens in the watery medium so that it no longer lies in the line of sight. This method was known in India in the prehistoric era and is still used in some parts of the world.[2] Cheselden also used this method. The immediate effects of couching are usually good, but the operation is often followed by infection which makes the eye useless once again. Another disadvantage is that even after an operation carried out appropriately the opaque lens can return to where it was so that any advantage initially gained is lost.

In order to avoid these problems a new method was developed, known as *extraction* of the cataract.[3] In this operation the doctor first makes an opening in the cornea and then pushes a tiny spatula through the anterior chamber and the pupil until he reaches the upper part of the cataract; he then opens the lens capsule and removes the opaque lens tissue piece by piece viâ the opening in the cornea. Jean Méry (1645–1722) was the first to see the

[1] A healthy lens is transparent thanks to the regular placing of the lens tissue fibres. The formation of deviant fibres results in a more or less opaque lens.

[2] Susruta (1000 B.C.?) may have been the first surgeon who performed the couching operation (see Bidyadhar 1939, 1940, 1941, and Dutt 1938). The first description of a cataract operation in European literature can be found in Celsus's *De re medicina* (first half of the first century) (Celsus 1935–1938, vol. 3, pp. 349–353). It has been suggested that Jesus effected his miraculous cures of blind people (see, for instance, Mark 8: 22–26) by pushing aside the opaque lenses of cataract patients (see Forrest 1965 and Forrest 1974). (Cheselden and the Biblical miracles were already mentioned in one breath in Charles Bew 1800, p. 58.)

[3] Ammar extracted opaque lenses with hollow needles as early as 1000 A.D., but this had been forgotten. See Hirschberg 1905, pp. 1128–1129.

advantages offered by this method.[1] In the *Histoire de l'Académie Royale des Sciences* of 1707 we read that

> M. Méry croit qu'on pourroit tirer les Cataractes hors de l'œil par une incision faite à la Cornée, & que cette maniere dont il ne paroît pas qu'il y ait rien à apprehender, préviendroit tous les périls ou les inconveniences de l'operation ordinaire. Il est bien sûr que la Cataracte ne remonteroit point, & ne causeroit point les inflammations qu'elle peut causer, lorsqu'on la loge par force dans le bas de l'œil.[2]

A third advantage of the new method was that it was no longer necessary to postpone the operation until the cataract was mature or hard.[3]

The first successful cataract extraction was performed by the Parisian doctor Petit on 17 April 1708.[4] Despite the advantages of the new method, doctors continued to use the old one until Jacques Daviel, "oculiste ordinaire de Louis XV," improved Petit's method and instruments in 1753.[5] This was the reason why Daviel is usually said to be the discoverer of the cataract extraction technique. In the second half of the eighteenth century the advantages and disadvantages of reclination and extraction were the subject of heated debate.[6]

It is important to note here that eyesight can never be perfect after these two types of operation since the lens has disappeared and, with it, part of the eye's refractive capacity. A cataract operation makes the patient far-sighted and if good eyesight is to be restored there is need of optical correction, in

---

[1] Méry was present at the eye operation carried out by Charles de Saint-Yves on Sunday 20 February 1707 on a patient whose lens was pressing against the iris. Saint-Yves removed the lens from the eye, thus performing the first recorded lens extraction in modern times. See Münchow 1984, p. 264.

[2] [Fontenelle] 1707, p. 24. I do not know whether Münchow's (1984, p. 280) ascription of this anonymous article to Fontenelle is correct.

[3] It has been discovered that it is important that babies suffering from double cataract should be operated on as soon as possible (preferably before they are two months old) since otherwise they will develop amblyopia as a result of visual deprivation.

[4] Petit performed the operation after consultation with and in the presence of Méry and Saint-Yves, who had previously carried out a lens extraction. See Münchow 1984, pp. 264–265.

[5] In 1747 Daviel (1693–1762) performed his first cataract extraction after an attempt at couching had failed. From 1750 onwards he decided to use the new method exclusively. He published his findings in Daviel 1748 and Daviel 1753.

[6] Daviel's method was fiercely supported by, for example, the elder Baron Wenzel—who did, however, admit to having destroyed "a hatful of eyes" before mastering the technique (Lebensohn 1969, p. 176). Gerard ten Haaff (1720–1791), lector at the *Illustere School* of Rotterdam, was the first surgeon in the United Republic of the Netherlands to apply Daviel's method. See Ten Haaff 1761 and Henkes 1982, p. 11.

the form of a lens or glasses.[1] The need for optical correction was already perceived by Benito Daza de Valdes in 1623. In his *Uso de los antojos*, one of the first books about spectacles, he recommended the wearing of glasses for patients who had had a cataract operation. Similarly, Jacques Rohault wrote in his *Traité de physique* (1672): "ceux à qui l'on a osté des Cataractes ne sçauroient voir que confusément; [...] ils ont besoin de lunettes fort pour voir distinctement."[2] And Louis wrote in the *Encyclopédie*: "après qu'on a abaissé la *cataracte*, la personne ne peut plus voir qu'à l'aide d'une verre lenticulaire."[3] As will be seen in the next few sections, this fact was missed by all the eighteenth-century commentators on Cheselden's report.

4 Cheselden's Observations

Many philosophers writing after 1728 on Molyneux's problem referred to Cheselden's report. Some of them agreed vociferously with Cheselden's results, while others criticised the report on methodological grounds. Moreover appreciation of the report was closely bound up with the position taken up by the commentator with regard to Berkeley's theory of vision. Those who rejected Berkeley's notions were usually critical of the report. But those who accepted Berkeley's ideas usually regarded Cheselden's observations as supportive of those ideas. This section will examine which observations were regarded as proof of the correctness of a negative answer to Molyneux's question.

As already said, Berkeley had stated that his theory of vision (and the solution he offered to Molyneux's problem) could be tested against the experiences of persons born blind receiving their sight through surgery. As early as 1710 he had pointed to Grant's eye operation and in *The Theory of Vision Vindicated and Explained* (1733) he summarised the most significant points in Cheselden's report. The fact that at the beginning the young man could not estimate distance and could not distinguish objects one from another, regardless of their size or shape, was regarded by Berkeley as definitive confirmation of his theory: "Thus, by fact and experiment, those points of the theory which seem the most remote from common apprehension were not a

[1] After a cataract operation the patient will normally need glasses of approximately +10 dioptres with a 2-dioptre cylinder at 180 degrees (to compensate for the vertical flattening of the cornea caused by the incision). These days it is also possible to fit a contact lens or an intra-ocular lens. See Trevor-Roper & Curran [1974] 1984, p. 452.

[2] Rohault 1672, p. 488.

[3] Louis 1752, p. 770.

little confirmed, many years after I had been led into the discovery of them by reasoning."[1]

Robert Smith (1689–1768), mathematician and professor of astronomy and experimental philosophy at Cambridge's Trinity College, quoted Cheselden's report in its entirety in his influential book *A Compleat System of Opticks* (1738).[2] This work, which deals with popular, mathematical, mechanical and philosophical aspects of optics, contains a chapter on the ideas that we obtain from our sense of sight and it was here that Smith described Molyneux's problem. As far as Smith was concerned, the answer given by Molyneux and Locke had turned out to be correct: "this opinion has since been confirmed by the experience of several persons, who receiving their sight from the operation of Couching, could not know any one thing from another, however different in shape and magnitude."[3] Smith referred to *The Tatler*[4] and to Cheselden's report. Smith suspected that the patient had learnt to recognise the position, size, shape and distance of objects by sight by following the movements of his hand with his eyes. The ideas acquired by sight would then, he believed, be associated with those gained through touch by means of habituation.

Berkeley's theory of vision gained a great deal of support especially in Great Britain, but there were also some followers to be found on the Continent. Voltaire (1694–1778), for instance, who visited England in the years 1726–1728, was impressed not only by the liberal political and cultural climate he found there but also by the experimental science of Newton, the empiricist philosophy of Locke and the ideas advanced by Berkeley regarding the sense of sight. The works he wrote after returning home included the *Elémens de la philosophie de Neuton* (1738), a popularisation of Newton's philosophy. The book contained not only an explanation of the way in which we learn to see such things as distance but also a summary of Cheselden's report, thereby making it accessible to the French-speaking world.[5]

Voltaire believed that distance, position, size and shape are not directly seen with the aid of optical lines and angles. Like Berkeley he considered that this had been proved by the reactions of people born blind who had undergone surgery:

> Tout cela ne pouvait être éclairci et mis hors de toute contestation que par quelque aveugle-né à qui on aurait donné le sens de la vue. Car si cet aveugle, au moment qu'il eût ouvert les yeux, eût jugé

---

[1] Berkeley [1733] 1975, §71.

[2] Smith's *Opticks*, which earned him the nickname "Old Focus," was translated into Dutch in 1753, into German in 1755 and into French in 1767.

[3] Smith 1738, bk. I, ch. v, art. 132.

[4] [Steele] [1709] 1898.

[5] Voltaire incorrectly dated Cheselden's report 1729.

> des distances, des grandeurs et des situations, il eût été vrai que les angles optiques, formés tout d'un coup dans sa rétine, eussent été les causes immédiates de ses sentimens. [...] Mais où trouver l'aveugle dont dépendait la décision indubitable de cette question?[1]

As could have been anticipated, Voltaire found this blind person in Cheselden's patient. Voltaire provided a summary of the report written by this "clever and skilful" surgeon and invented a few details himself. He wrote, for example, that at the beginning the young man could not distinguish between objects that he had assessed, using his hands, as rounded or angled.[2] Voltaire was convinced that the experiences of the young man confirmed everything that Locke and Berkeley had so accurately predicted: "Il ne distingua de long-temps ni grandeurs, ni situations, ni figures même."[3]

Voltaire considered his opinion regarding how we do *not* see things to have been justified. The question now was how we *do* imagine shapes and suchlike. His answer, which betrays Berkeley's influence, was as follows: in the same way as we imagine people's passions, namely by reading their facial expressions and the colour of their faces. It is a language spoken by nature to all eyes; but we need *experience* to learn to understand it. Voltaire compared the way in which we learn to see with the way in which we learn to speak or read, with the difference that seeing is easier.

Strictly speaking, said Voltaire, distance, size and position are not visible things; the real and direct object of seeing is nothing more than coloured light. All the rest we observe only after a time and through experience. And since as far as the representation is concerned all mankind has the same language which apparently links colour to objects, we could be tempted to believe that we *see* extension. We could also tend to think that there is a necessary link between words and meanings if everyone spoke the same language, but we know that such is not the case.

Voltaire believed that in addition to experience we need the help of other senses in order to assess a (sudden and arbitrary) judgment about distance and suchlike. But he tempered this opinion by stating that a being that possessed no more than the sense of sight would still be able to form ideas of distance and such through experience:

> Sans doute ses idées d'étendue, de distance, ne seraient pas rigoureusement les mêmes que les nôtres, puisque le sens du toucher n'aurait

[1] Voltaire [1738] 1819, part II, ch. vii.

[2] This reminds us of Molyneux's question, but Voltaire did not mention that problem explicitly.

[3] Voltaire [1738] 1819, part II, ch. vii. A similar opinion was expressed by (among others) Le Cat [1740] 1744, Krüger 1756, Morand 1757, Condorcet [1773] 1847, Von Baer 1824, and Apelt 1857.

> pas contribué à les former: sans doute ses jugemens sur le lieu, la forme, la distance, seraient plus souvent erronés que les nôtres, parce qu'il n'aurait pu les rectifier par le toucher. Mais il est très probable que c'est à quoi se bornerait toute la différence entre lui et nous.[1]

How this creature would form these ideas according to Voltaire is not clear. But the strong suspicion expressed by Voltaire in this passage would certainly not have been confirmed by Berkeley who, in fact, believed that an *unbodied spirit* would be unable to perceive extension or shapes.

Whereas Voltaire introduced Cheselden's report into the French-speaking world, it was Willem Jacob 's Gravesande who was the first Dutchman to allude to it, in his *Introductio ad philosophiam, metaphysicam et logicam continens* (1736).[2] The relevant passage reads as follows (in the second edition of 1737):

> Facultatem hanc de objectis visis judicandi, Usui & Experientiæ tantum deberi, ex natura rei deducitur. Sed omnis scrupulus removetur, si perpendatur historia Juvenis cœci nati, & qui, inter annum ætatis decimum tertium & decimum quartum, visum recuperavit; quæ invenitur in Actis Philosophicis Societatis Regiæ Anglicanæ. Num. 402. Art. 7.[3]

Like his colleague Petrus van Musschenbroek, 's Gravesande had been influenced by the new experimental philosophy during a visit to England.[4] Both imitated Locke in fiercely opposing the notion that we have inborn ideas of external objects. They were convinced that people lacking one of the senses would not be capable of forming ideas of the related characteristics of objects. Both philosophers referred to cured cataract patients in order to support their belief that we *learn* to see.[5]

---

[1] Voltaire [1738] 1819, part II, ch. vii, note 13.

[2] 's Gravesande (1688–1742), professor of mathematics, astronomy and philosophy, was the first person in mainland Europe to teach experimental physics in the spirit of Newton.

[3] 's Gravesande [1736] 1737, §509. Translation: "The fact that we are able to make judgments about the distances of objects seen on the basis of Habit and Experience alone is deduced from the nature of the matter itself; but all objections are removed when we consider the account of a Young Man blind from birth who between the thirteenth and fourteenth year of his life received the power of Sight again; which account is to be found in the Physical Accounts of the Royal Society in England, No. 402, Art. 7."

[4] Van Musschenbroek (1692–1761) was named professor of mathematics and philosophy in Duisburg in 1719 and in Leiden in 1740. See De Pater 1979 and De Pater 1988.

[5] Van Musschenbroek 1762, §1889, compares children and cataract patients who had been operated on with people who have never seen an object through a microscope or telescope: at first they will not be able to see any object at all, but the more frequently they use the instrument the better they will be able to distinguish the various objects one from the other.

's Gravesande and Van Musschenbroek made no comment on Molyneux's problem. However, one of their disciples, the versatile scholar Petrus Camper (1722–1789) did.[1] From being a student Camper was interested in descriptions and explanations of the faculty of sight and, later, eye diseases. His two dissertations witness to this: *Dissertatio optica de visu* (1746) and *Dissertatio physiologica quibusdam oculi partibus* (1746);[2] so also his book *De oculorum fabrica et morbis* (1766).[3] Camper operated on cataract sufferers himself; he was a supporter of Daviel's new method, extraction of the cataract.

Camper discussed Cheselden's report and Molyneux's problem in his *Dissertatio optica de visu* (1746). He stated that no-one doubted the fact that we can distinguish the form of objects by sight alone. However, we would never have acquired the idea of the shape of an object, he continued, if we had not already explored it by touch. He reasoned as follows: the stimulation of a nerve and the corresponding perception have nothing in common and therefore neither do the perception and the object perceived have anything in common. With the exception of touch, none of our senses teach us anything *per se.* For when we touch an object we are immediately aware of its resistance and shape. Later, when we have learnt these characteristics through touch, we begin to pay attention to the circumstances. Then we see that objects are illuminated in various ways and we relate that to what we have felt. And thus we learn to distinguish different objects, such as a sphere and a cube.[4] Camper was of the opinion that Cheselden's report proved that Molyneux and Locke had provided the correct answer to Molyneux's problem: the young man was well able to recognise the shape of objects by touch but not by sight.

Although there are more philosophers who referred to Cheselden's observations in support of their negative answer to Molyneux's question and in support of Berkeley's theory of vision, we will limit ourselves to those already mentioned. What was seen as by far the most important fact relative to Molyneux's problem was that Cheselden's patient had not been able to distinguish objects from one another, no matter whether they differed as regards shape and size. On this basis they concluded that a person born blind and restored to sight was unable to distinguish a sphere and a cube from one another and to name them.

---

[1] Camper's interests included ophthalmology, midwifery, anatomy, veterinary medicine, mineralogy, palæontology and economics. He was a skilled draughtsman and instrument maker. See Rochat 1939 and Ten Doesschate's "Introduction" in Camper [1746] 1962. Goethe called Camper "ein Meteor von Geist, Wissenschaft, Talent und Thätigkeit"; quoted in the *Nieuw Nederlandsch biografisch woordenboek* (1911) part 1, p. 555.

[2] Camper defended these two dissertations on the same day.

[3] This book was not published until 1913.

[4] Camper [1746] 1962, ch. II, §i.

In the end the unconditional acceptance of Cheselden's report failed to lead to any new insights into Molyneux's problem. The criticism aimed at the report did, however, raise some interesting points of view.

5 The Methodology of Psychological Experiments

The methods employed in and the results obtained from Cheselden's studies were challenged mainly in France, where Berkeley's teachings on the faculty of sight turned out to be less sacrosanct than in Great Britain.[1] A number of *philosophes* wondered, for instance, if the patient's eyes immediately following the operation had been in good condition and whether the patient had been questioned in an appropriate, non-suggestive manner. Julien Offray de La Mettrie (1709–1751) was the first to attempt to deal with these matters.

La Mettrie, whose reputation was largely due to his *L'homme machine* (1747), had previously proclaimed his materialistic doctrine anonymously in the *Histoire naturelle de l'âme* (1745).[2] As is well known, La Mettrie opposed the dualism of body and soul as proclaimed by Descartes. La Mettrie assumed that the activities of the soul are dependent on the structure and functions of the nervous system. Like the body, therefore, the soul would be amenable to experimental research. The "natural history of the soul" should therefore not be the territory of the theologian or metaphysician but of the natural scientist. La Mettrie deprived mankind of the status that Descartes still accorded the species. Where Descartes excluded mankind from his notion of all living creatures as automata, La Mettrie regarded man too as a machine or self-regulating system. Since in La Mettrie's view nature was self-explanatory, he did not require a hypothesis of a creator God. As could be expected, La Mettrie's materialistic-atheist philosophy met with fierce resistance, but it also had considerable influence.

In the *Histoire naturelle de l'âme* La Mettrie revealed himself a defender of the notion that all our ideas come from our senses. At the end of the book he set out a number of true accounts designed to support this statement.[3] They are stories of individuals whose fate differed from that of "normal" people. The first concerned a deaf-and-dumb boy from Chartres who suddenly

[1] In Great Britain criticism of Berkeley's theory of vision and his solution to Molyneux's problem came from such quarters as Anti-Berkeley 1752 and Porterfield 1759.

[2] The title page was designed to give the impression that the book had been translated from English. The work was reprinted (with modifications) as *Traité de l'âme* in La Mettrie's *Œuvres philosophiques* (La Mettrie 1751).

[3] La Mettrie 1745, ch. XVII; this is chapter XV in La Mettrie 1751.

began to speak one day.[1] From what he said it became apparent that he had absolutely no notion of God, the soul, goodness or death. Although he had always attended Mass, crossed himself and knelt, he had no idea what it all meant. La Mettrie believed that this indicated that ideas of God and suchlike are not inborn. Another story concerned a young boy who in 1694 was discovered in the forests on the Russian-Lithuanian border living in the midst of a group of bears.[2] The boy looked appalling, walked on all fours, produced rough sounds and did not use his intelligence. Once he had been reintroduced to his own kind and had learnt to speak a few words, it turned out that he had no memory of his former circumstances. Both stories were recounted by other Enlightened thinkers, partly to demonstrate the correctness of empiricism but also because of their curiosity value.

La Mettrie also recounted the suspicion expressed by Arnobius that a child brought up in a cave with little to stimulate him would remain as dumb as an ox. He would have no idea of plants, animals, the earth and the sea, nor of metaphysics, morality or mathematics.[3] La Mettrie also described Amman's method of teaching deaf-and-dumb people to speak, underlined the importance of good upbringing, told tales of wild men and satyrs and quoted the story of Cheselden's blind boy.[4] All these stories, La Mettrie believed, proved his book's conclusion: "Point de sens, point d'idées. Moins on a de sens, moins on a d'idées. Peu d'éducation, peu d'idées. Point de sensations reçües, point d'idées."[5]

La Mettrie stated (not entirely correctly) that Voltaire had written that Cheselden's patient had only perceived coloured light immediately following the operation, without being able to distinguish a sphere from a cube and without any idea of extension, distance, shape, etc.[6] However, La Mettrie believed that a person born blind and recovering the faculty of sight would be able to distinguish different objects from one another:

> un globe attentivement considéré par le toucher, clairement imaginé & conçu, n'à qu'à se montrer aux yeux ouverts; il sera conforme à

[1] The original text dealing with the deaf boy of Chartres can be found in the *Histoire de l'Académie des Sciences pour l'année* 1703, second edition (Paris 1720), pp. 18–19.

[2] According to Verbeek 1988, vol. 1, p. 117, La Mettrie took this story from Bernard Connor 1697.

[3] Arnobius, *Adversus gentes* (c. 300 A.D.), bk. II. See Arnobius 1949 for a modern translation.

[4] The physician J. C. Amman (1669–1724) occupied himself with the education of deaf-and-dumb people. See Amman 1692.

[5] La Mettrie 1745, "Conclusion de l'ouvrage."

[6] La Mettrie was the first to point out that Cheselden's account was inconsistent: the boy was said not to perceive any size, and yet he said that his thumb was as big as a house. La Mettrie 1745, ch. XVII, *Histoire* iii.

> l'image, ou à l'idée gravée dans le cerveau; & conséquemment il ne sera pas possible à l'Âme de ne pas distinguer cette figure de toute autre, si l'organe dioptrique a l'arrangement interne nécessaire à la vision.[1]

La Mettrie also considered it unimaginable that a skilful anatomist would not be able to recognise all the bones of the human body by touch and make a skeleton of them. Because he would recognise by touch that which he had seen: "Les idées reçües par les yeux se retrouvent en touchant, & celles du tact, en voyant."[2] This was the position usually adopted by rationalist and not so much by empiricist thinkers.

As far as Cheselden's report was concerned, La Mettrie believed that a prejudice had been created by Locke's answer to Molyneux's question. A possible explanation of the results of the report was, thought La Mettrie, that the eye had not yet recovered from the operation or that the young man had been influenced:

> Ou on n'a pas donné le tems à l'organe dioptrique ébranlé, de se remettre dans son assiéte naturelle; ou à force de tourmenter le nouveau voyant, on lui a fait dire ce qu'on étoit bien aise qu'il dît.[3]

Because, stated La Mettrie, people were more interested in finding support for errors than in discovering the truth. The skilful theologians who had questioned the deaf boy of Chartres expected to find in the nature of man judgments which would precede the first perception. But God, who according to La Mettrie does nothing that is useless, has not even given us a single first idea of his own attributes. And thus inborn ideas of shape and space would have been useless for Cheselden's patient in enabling him to differentiate at first sight between a cube and a sphere. He would only require time to open his eyes and observe the richly endowed theatre of the universe.

By drawing attention to the physiological condition of the eyes immediately following the operation and to the nature of the questioning of the patient La Mettrie gave a new boost to the discussions surrounding Molyneux's question.

The Frenchman Étienne Bonnot de Condillac (1714–1780) also criticised Cheselden's report. His first riposte was contained in his *Essai sur l'origine des connoissances humaines* (1746). The motivation behind Condillac's *Essai* was a desire to improve on Locke's *Essay.* Condillac admired Locke in that the latter had pointed to experience as the source of all knowledge and had

[1] La Mettrie 1745, ch. XVII, *Histoire* iii.
[2] La Mettrie 1745, ch. XVII, *Histoire* iii.
[3] La Mettrie 1745, ch. XVII, *Histoire* iii.

opened up new paths by his observation and analysis of the human mind. But Condillac accused Locke of not having succeeded in setting out a consistent form of empiricism since he had introduced reflection as an inborn capacity. Condillac's *Essai* was intended to correct Locke's failures and it also contained the first criticism of Berkeley's theory of vision.

Condillac's entire philosophy was based on the statement that we may not ascribe any capacities to the mind of which we are not *aware* or of which we cannot make ourselves aware. He thus criticised both Locke and Berkeley who had assumed unconscious judgments as an explanation of certain phenomena of perception. Judgments of this type, said Condillac, would be in conflict with our experience. Molyneux's problem, which Locke had introduced in order to illustrate his thesis, would in no way refute this.[1]

Condillac agreed with Locke that the retinal image of a sphere was an unevenly coloured circle. But he did not accept that the sensation we have by means of the retinal image is the sensation of a circle:

> Parmi ces suppositions, Locke avance, sans preuve, que la sensation de l'âme ne représente rien de plus que l'image que nous savons se tracer dans l'œil. Pour moi, quand je regarde un globe, je vois autre chose qu'un cercle plat: expérience à laquelle il me paroît tout naturel de m'en rapporter.[2]

In Condillac's opinion Locke should have been consistent and should have argued about distance, position, size and extension in the same way that he had concerning shape. But Locke had not done so: in Molyneux's problem he required the globe and the cube to be of approximately the same size, thereby implying that sight can give us various notions of size without the help of any other judgment whatsoever. Locke thus contradicted himself, believed Condillac, since it is impossible to understand how one can have ideas of size without having ideas of shape.

In contrast to Locke, Berkeley had drawn the conclusion that a blind person suddenly being given the faculty of sight would be unable to determine not only shape, but also position, distance and size. Berkeley had invented a thought experiment regarding an *unbodied spirit* in order to demonstrate that the pure objects of sight cannot be considered geometrical forms. This kind of *œil animé*, as Condillac called it, would in Berkeley's view be able to see only coloured light, but no extension etc. It would become accustomed, stated Condillac, to judging the whole of nature as a mathematical point. But what we in fact see, according to Condillac, is light and colours which of necessity delineate different distances, different sizes and different positions.

---

[1] Condillac [1746] 1947, part I, ch. vi, §1.
[2] Condillac [1746] 1947, part I, ch. vi, §3.

As far as Molyneux's problem was concerned, Condillac believed that a person cured of congenital blindness would not immediately be able to enjoy the spectacle offered to him by light and colour:

> il ne faut pas croire qu'au moment qu'il ouvre les yeux, il jouisse déjà du spectacle que produit dans toute la nature ce mélange admirable de lumière et de couleur. C'est un trésor qui est renfermé dans les nouvelles sensations qu'il éprouve; la réflexion peut seule le lui découvrir et lui en donner la vraie jouissance.[1]

When we ourselves turn our eyes to a richly decorated painting and see it in its entirety we do not as yet form any particular idea of it. We must first, said Condillac, consider all its parts separately.

Once the person born blind is capable of reflecting on what he sees, Condillac believed that he would no longer see a point but extension with a certain length, breadth and depth. When he comes to analyzing this extension, he will form ideas for himself of surface, line, point and all kinds of shapes, and those ideas will be identical to those which he has acquired through touch:

> car, de quelque sens que l'étendue vienne à notre connoissance, elle ne peut être représentée de deux manières différentes. Que je voie ou que je touche un cercle et une règle, l'idée de l'un ne peut jamais offrir qu'une ligne courbe, et celle de l'autre qu'une ligne droite.[2]

The person born blind would thus be able to distinguish the sphere from the cube by sight, because he would recognise in them the same ideas he had formed through the sense of touch.[3]

It would be possible, said Condillac, to have the man defer his judgment by asking him what made him so sure that bodies must have the same shape when perceived by sight as when perceived by touch. Indeed, it could well be that an object that looks like a sphere may appear on touching it to be a cube. Condillac was of the opinion that it was awkward to find a convincing proof of this; only experience would provide a solution.[4]

Condillac admitted that Cheselden's report seemed to create difficulties for him: "c'est une expérience qui paroît, en tous points, contraire au sentiment que je viens d'établir."[5] He explained the apparent contradictions, as

---

[1] Condillac [1746] 1947, part I, ch. vi, §14.

[2] Condillac [1746] 1947, part I, ch. vi, §14.

[3] Condillac [1746] 1947, part I, ch. vi, §14.

[4] Condillac [1746] 1947, part I, ch. vi, §14. See also Thomson 1974 and Pitcher 1974.

[5] Condillac [1746] 1947, part I, ch. vi, §15. Condillac quoted Voltaire's version of Cheselden's report.

La Mettrie had done, by pointing out that the young man's eyes did not function well immediately after the operation because they had not been used for years. He would require many days of exercise before being able to persuade the different parts of the eye to work together in a coordinated fashion after having become stiffened in the course of time. And, said Condillac, that was the reason why Cheselden's patient had subsequently gone round feeling his way in the dark for another two months. In contrast to Berkeley, Condillac believed that if the young man had not used his hands he would have obtained by the power of sight the same ideas as those which he had earlier obtained through touch. The only difference would be that it would have taken longer.

Condillac accused of bias those who could think of no other explanation for the young man's weak powers of sight than those advanced by Locke and Berkeley. And, he added, they were the wrong reasons since they were not in agreement with the principle that only that may be proposed which is indisputable and which everyone with the least powers of reflection can note in themselves.

Where Condillac defended in his *Essai* the thesis that from birth we can perceive size, shape, distance and position immediately outside ourselves, Mlle Élisabeth Ferrand (1700–1752), "the Egeria of the Enlightenment," was able to convince Condillac of the incorrectness of his thesis.[1] This caused him to change his view radically: in his later work, the *Traité des sensations* (1754), he praised Berkeley's insights and qualified his own earlier opinions as prejudice.

The main cause of the widespread belief in this prejudice was, believed Condillac, that we have formed such a deeply rooted habit of judging from sight the things with which we are surrounded that we can no longer remember what we saw when we first opened our eyes. Molyneux had proposed his problem of the man born blind for various reasons, including a desire to investigate what we see by nature, without being prejudiced by habit. Condillac wanted to do something similar, but for all the senses. He wanted a tool with which we could more easily imagine what we owe to each sense. The tool was handed to him by Mlle Ferrand. She suggested that Condillac should imagine a statue whose various senses should be opened up. In fact it was more a question of a series of statues: a statue that was open exclusively to the sense of smell (or taste or sight, etc.); a statue that disposed of the sense of smell plus taste (or smell plus sight, etc.), etc.

In fact Condillac was not the only one to set up a thought experiment of this type. Denis Diderot had already suggested, in his *Lettre sur les aveugles* (1749), that it should be possible to imagine a block of marble able to

[1] See Bongie 1977, p. 150, and Bongie 1978, p. 92, note 34.

think and feel.[1] In his *Lettre sur les sourds et muets* (1751) he had written that he wanted to conduct a sort of metaphysical anatomy by dividing up an individual, as it were, into five persons, each of which would have only one sense.[2] And Buffon had wondered what the perceptions and judgments would be of the first man at the instant of creation.[3] On these grounds Condillac was accused of plagiarism.[4] The force of his thought experiment was also the subject of discussion. Grimm, for instance, was of the opinion that Condillac's assumptions were arbitrary and impossible and that it would be better to "faire la véritable histoire métaphysique de l'homme."[5]

Condillac believed that a statue that was endowed exclusively with the sense of sight would only have sensations of light and colour and would not realise that such sensations were in fact caused by external objects. If the statue were given the additional sense of touch, it would also be able to perceive size, shape, distance and position. If it were to see only those objects which it touched and touch only those objects which it saw, it would be unable to distinguish the sensations of sight from those of touch. Since our power of sight always works jointly with the sense of touch, the sensations of sight would mingle with those of touch.[6]

Condillac emphasised that there is a difference between seeing (*voir*) and looking (*regarder*); certainly the statue would see immediately, but it would have to learn to look and to know what it saw. In contrast to Locke, Condillac denied the existence of unconscious judgments. We do not see objects first as flat bodies and then as solid, stated Condillac, but it is more a case of the experience of touch being necessary for us to realise that we are seeing objects.

According to Condillac, Berkeley was the first to note that the eye cannot of itself judge distance, position, shape and size. The man born blind in Moly-

---

[1] Diderot [1749] 1961, p. 144: "Madame, combien nos sens nous suggèrent de choses; et que nous aurions de peine, sans nos yeux, à supposer qu'un bloc de marbre ne pense ni ne sent!"

[2] Diderot believed that these five persons would be able to communicate with one another using geometry: "par la faculté qu'elles auroient d'abstraire, elles pourroient toutes être géomètres, s'entendre à merveille, & ne s'entendre qu'en géométrie." Diderot [1751] 1965, p. 45.

[3] Buffon 1749–1789, vol. 3 (1749), *Histoire naturelle de l'homme*, chapter "Des sens en général."

[4] See, for instance, F. M. Grimm, "Lettre 1er novembre 1755," in Grimm & Diderot 1813, vol. 1, p. 445: "M. l'abbé Condillac avoit noyé la statue de M. de Buffon dans un tonneau d'eau froide." Condillac defended himself against the accusations in his "Réponse à un reproche," which was appended to his *Traité des sensations*.

[5] Grimm, "Lettre 1er décembre 1754," in Grimm & Diderot 1813, vol. 1, p. 263.

[6] Condillac [1754] 1947, part III, ch. iv, §2: "ses sensations se mêlent avec les idées qu'elle lui doit."

neux's problem would be unable to do any of this; he would be unable, using his sight alone, to distinguish a sphere from a cube.[1] Condillac, incidentally, believed that the conditions that the sphere and the cube should be made of the same material and should be the same size were superfluous.[2]

In the *Traité des sensations* Condillac wrote that he perceived a proof of Berkeley's thesis in Cheselden's report, but he also had a critical remark to make about it.[3] Thus he found that Cheselden had not always expressed himself sufficiently carefully. He had written, on the one hand, that the young man started off by being unable to distinguish objects one from the other, no matter how different they were in shape or size; on the other hand he noted that the young man found that regular objects were the most pleasant. However, Condillac did not blame him for this: he regarded Cheselden as a pioneer, seeing for the first time phenomena consisting of "thousands of details difficult to comprehend." In such cases there is always a great deal left to be desired. But such matters should at the very least provide insight, so that observations could be made with a greater degree of success on a subsequent occasion. In order to avoid the same mistakes at a later date Condillac followed Diderot's example in making some recommendations as to how a blind person operated on to restore his sight could best be observed.[4]

Condillac recommended first of all that one must ensure that the person born blind should reflect on the ideas that he has obtained through the sense of touch so that later he can of himself (without others having to ask him questions) say what ideas he obtains through the faculty of sight. Once the operation for cataract has been performed the patient should be forbidden to touch anything until he has learnt what ideas belong to the faculty of sight. Investigations should be carried out to ascertain whether light appears to him to be extended or whether he can determine its limits and whether it is so vague that he cannot observe variations in it.

The patient should then be shown two colours, first separately and then simultaneously, asking him if there is anything he recognises as having seen previously. This should be repeated with an ever-increasing number of colours. And investigations should above all be made to discover whether he can deter-

[1] Condillac [1754] 1947, part III, ch. iv, §3. Bonnet 1760 also uses the notion of a statue. Bonnet was of the opinion that a blind person who had been operated on would not be able to recognise a round object by sight since there would be absolutely no link between tactile and visual ideas.

[2] Condillac [1754] 1947, part III, ch. iv, §3.

[3] Condillac [1754] 1947, part III, ch. v. In 1754 Condillac discussed Cheselden's report in far greater detail than he had done in 1746. It is likely that he was taking his data no longer exclusively from Voltaire's *Élemens* (1738) but also from Buffon 1749–1789, vol. 3 (1749), *Histoire naturelle de l'homme*, chapter "Du sens de la vue." See Pastore 1973.

[4] Condillac [1754] 1947, part III, ch. vi.

mine size, shape, position, distance and movement. He should be questioned skilfully, all leading questions being avoided. After all, asking him if he can see a triangle or a rectangle would mean that he was being told what he should see. And thus Condillac implicitly characterised Molyneux's question as a bad question and a leading question.

Condillac invented a somewhat remarkable means of removing all shadow of doubt. A blind person just operated on should be closed up in a glass room. This would lead to two possibilities: either the patient will recognise the objects situated on the other side of the glass and will determine their shape and size; or he will merely note the space enclosed by the walls of the room and will interpret the objects as surfaces of different colours, which appear to extend as far as he can move his hand towards them. The first case, said Condillac, would prove that the eye can judge without assistance from touch; the second that it can only judge after having consulted touch. Condillac suspected that the latter would be the case.[1]

Denis Diderot (1713–1784), who was Condillac's contemporary, distinguished himself from his predecessors because his interest in those born blind was not limited to phenomena in the field of epistemology or the psychology of perception but concerned their whole world. In his wonderful *Lettre sur les aveugles, à l'usage de ceux qui voient* (1749) he described how greatly the world of the blind differs from that of the sighted.[2] He used as a basis a discussion he had had with an educated person from Puiseaux, blind from birth, and on statements made by the blind mathematician, Saunderson, already referred to here. They proved that the ideas the blind have regarding beauty differ from those held by the sighted and, in addition, that the blind have different morals and metaphysics. They thereby confirmed a more general suspicion on the part of Diderot:

> je n'ai jamais douté que l'état de nos organes et de nos sens n'ait pas beaucoup d'influence sur notre métaphysique et sur notre morale, et que nos idées les plus purement intellectuelles, si je puis parler ainsi, ne tiennent de fort près à la conformation de notre corps.[3]

One could also state that the religious opinions and attitudes of blind people are vastly different from those entertained by the sighted. After all, the blind are not capable of witnessing the wonders of nature with their own eyes; and

---

[1] Condillac [1754] 1947, part III, ch. vi, §2.

[2] Diderot's *Lettre sur les aveugles* was published anonymously. Together with Molyneux's problem and Cheselden's report the letter was extensively discussed in an article by D'Alembert in the *Encyclopédie* (D'Alembert 1751). Paulson 1987 contains an interesting chapter about Diderot's letter. See also Perkins 1978.

[3] Diderot [1749] 1961, p. 92.

moreover, said Diderot, they are themselves a monstrous product of nature. Diderot put the following provocative words in Saunderson's mouth while the latter was on his deathbed and was talking to the Reverend Holmes:[1]

> ce beau spectacle qui n'a jamais été fait pour moi! J'ai été condamné à passer ma vie dans les ténèbres; et vous me citez des prodiges que je n'entends point, et qui ne prouvent que pour vous et pour ceux qui voient comme vous. Si vous voulez que je croie en Dieu, il faut que vous me le fassiez toucher.[2]

Diderot then allowed Saunderson to speculate on the origin of time and objects. Originally there would have been overall chaos, but this would have slowly resolved itself and a large number of misformed and a few well-formed beings would have come into existence. By natural selection the monsters would be destroyed in the course of time, thus producing the wondrous order so highly appreciated by Newton, Leibniz and Clarke. But the order in the world is not perfect, remarked Saunderson, pointing to himself: "l'ordre n'est pas si parfait [...] qu'il ne paraisse encore de temps en temps des productions monstrueuses."[3]

The discussion with the Reverend Holmes made such a deep impression on Saunderson that he fell into a delirium, uttered the immortal line "*O Dieu de Clarke et de Newton, prends pitié de moi!*" and gave up the spirit.[4] The atheist attitude expressed by Diderot in the *Lettre sur les aveugles* was not appreciated: the French government locked him up in the prison of Vincennes.

A large part of the *Lettre sur les aveugles* deals with Molyneux's problem. Locke, Berkeley and many others had used Molyneux's thought experiment as a means of better reflecting on our original perceptions. By putting ourselves in the shoes of someone born blind it should be easier for us to imagine what we saw before we had formed judgments and habits. Diderot, however, did not see the sense in such an enterprise:

> On cherche à restituer la vue à des aveugles-nés; mais si l'on y regardait de plus près, on trouverait, je crois, qu'il y a bien autant à profiter pour la philosophie en questionnant un aveugle de bon sens.

[1] Holmes does indeed appear to have been a witness of Saunderson's death: "The reverend Gervas Holmes informed him [i.e., Saunderson] that the mortification gained so much ground that his best friends could entertain no hope of his recovery. He received this notice of his approaching death with great calmness and serenity and after a short silence, resumed life and spirits." Quoted by Vernière in Diderot [1749] 1961, p. 118, note 2. What Diderot wrote was, of course, apocryphal.

[2] Diderot [1749] 1961, pp. 118–119.

[3] Diderot [1749] 1961, p. 122.

[4] Diderot [1749] 1961, p. 124.

> On en apprendrait comment les choses se passent en lui; on les comparerait avec la manière dont elles se passent en nous, et l'on tirerait peut-être de cette comparaison la solution des difficultés qui rendent la théorie de la vision et des sens si embarrassée et si incertaine. [...] J'aurais moins de confiance dans les réponses d'une personne qui voit pour la première fois, que dans les découvertes d'un philosophe qui aurait bien médité son sujet dans l'obscurité, ou, pour vous parler le langage des poètes, qui se serait crevé les yeux pour connaître plus aisément comment se fait la vision.[1]

If, despite these difficulties, the decision is taken to operate on one born blind, said Diderot, a number of conditions need to be fulfilled in order to obtain reliable information.[2] First of all a blind person with a good dose of common sense is required, preferably a philosopher since such a person can think clearly. He would need to be thoroughly prepared and the observations should only begin some considerable time after the operation has been performed, when the eyes are completely healed. Meanwhile the patient should be kept in darkness, where he should be given the opportunity to exercise his eyes. Moreover—and this was a somewhat sticky problem—he should be questioned in a skilful manner so that he tells no more than that which is happening within him. Finally the questioning should take place before a forum of learned men.[3] In brief, "Préparer et interroger un aveugle-né n'eût point été une occupation indigne des talents réunis de Newton, Descartes, Locke et Leibniz."[4]

Like Boullier, Diderot replaced the sphere and the cube in Molyneux's problem with a *circle* and a *square*, since he believed that we can only judge distance from experience and that someone opening his eyes for the first time sees only (flat) surfaces and does not know that objects project forwards. And even if a person born blind perceived projections and solidity from the first instant of restored sight, and even if he could distinguish not only a circle from a square but also a sphere from a cube, he would not, asserted Diderot, be able to do this with more complex objects (such as a glove, a housecoat or a doctor's headgear).[5]

---

[1] Diderot [1749] 1961, pp. 126–127.

[2] See also Delacampagne 1987.

[3] At the beginning of his letter Diderot sarcastically accused doctor Réaumur of not allowing philosophers to be present at his eye operations.

[4] Diderot [1749] 1961, p. 128.

[5] Diderot [1749] 1961, p. 143. Diderot agreed with Condillac on the superfluous nature of the conditions that the sphere and the cube should be of the same size and made of the same material. Diderot provided an interesting variant of Molyneux's question: everyone agrees, he said, that touch does not teach sight to distinguish colours. If a person born

Diderot was of the opinion that Molyneux's question, when seen in a somewhat more general sense than Molyneux had done, actually contained two questions: (1) will the blind person be able to see immediately after the cataract operation? and (2) if so, will he see sufficiently well to be able to distinguish shapes; will he be capable, without hesitation, of giving the same names to objects by sight as he did by those of touch; and will he have a proof of the fact that he is using the correct names?

With regard to the first question, Diderot had no hesitation in saying that a blind person cured of cataract would, like a baby when it first opens its eyes, be incapable of seeing anything: "on n'est affecté, dans les premiers instants de la vision, que d'une multitude de sensations confuses."[1] Diderot believed that touch could help the eye to distinguish shapes one from the other. But he was also convinced that the eye would be capable of doing this without the aid of touch: "je pense nullement que l'œil ne puisse s'instruire, ou, s'il est permis de parler ainsi, s'expérimenter de lui-même."[2] A living eye (*un œil vivant et animé*) would, said Diderot, be able to distinguish the size and shape—or at least the rough outlines—of objects.

It in no way surprised Diderot that Cheselden's famous discoveries had concluded that his patient could distinguish nothing in the beginning. He concluded that the eye needed time to exercise and gain experience, and not that touch would be required to distinguish shapes. Moreover, added Diderot, what can we really expect of someone who was unaccustomed to reflect on himself and who was unaware of the advantages of sight, who was insensitive to his own unhappiness and was totally incapable of imagining the extent of the damage done to his own satisfaction by the loss of this sense? Saunderson would certainly not have been as indifferent as Cheselden's patient and he would most certainly have reacted differently.

If the person born blind is given the time, said Diderot, at some point he will succeed not only in distinguishing colours but also the overall outline of an object. Suppose that he masters this skill in a short time or that he should obtain it by moving his eyes in the dark: would he then be able to recognise and name objects by sight that he had previously touched? Diderot gave various answers to this question, depending on the intelligence of the blind person.[3]

---

blind should receive his sight and a ***black cube*** and a ***red sphere*** be placed against a white background, he would immediately distinguish the limits of these shapes. Diderot [1749] 1961, p. 137.

[1] Diderot [1749] 1961, p. 135.

[2] Diderot [1749] 1961, pp. 135–136.

[3] Diderot [1749] 1961, pp. 141–143.

Uncivilised people, without education, without knowledge and unprepared, will say that the one object is a circle and the other a square, without having any good grounds for making such a judgment. Or in their naïveté they will insist that in the objects presented to their sight they recognise nothing similar to anything they have ever touched. For in fact they are not used to any kind of reasoning at all. They do not know what sensations or ideas are and are incapable of comparing the representations obtained through touch with those received viâ the eyes. *Other people* will compare the shapes they see with shapes that have left an impression on their hands. They will mentally touch the objects that are at a distance and of the one they will say that it is a square and of the other that it is a circle—without knowing why.

A *metaphysician* will compare the ideas received through the eyes with those gained by touch. He will say that he is strongly tempted to believe that this is the object that he has always called a circle and the other the object that he has always called a square. But because he would not know whether what he sees is also perceivable by touch, the objects could be transformed under his hands and provide sensations by touch which would be opposite to those of sight. Thus he will be brought to confess that he has no knowledge whatsoever regarding the content of his statements. He will only say that this appears to him to be a circle and that a square.

If we replace the metaphysician with a *geometer*, or Locke with Saunderson, he will say that the one shape is a square and the other a circle. For he will note that only in the case of the first can he place threads and pins on his computing board to mark the four corners of a square, and only in the case of the second can he place the threads required to demonstrate the properties of a circle.[1] Unlike the metaphysician he will not say that what feels like a sphere perhaps looks like a cube. For he knows that those to whom he has demonstrated the properties of a circle in the past have not touched his instrumentarium and yet have understood him. The circle felt by the geometer agrees, therefore, with the circle seen by his audience.

Diderot's *Lettre sur les aveugles* represented an important contribution to the debate on Molyneux's problem. First of all he aroused discussion as to whether the perceptions of the blind person operated upon were appropriate to solve a problem dealing with sight. By asking this question he cast doubt on the basic assumptions of Molyneux, Berkeley and others. Secondly, like La Mettrie and Condillac before him, he pointed up the critical state in which the eye could be immediately following the operation. Diderot also drew a distinction between what someone opening his eyes for the first time would be

[1] In order to gain an impression of how Saunderson worked with his computing board, see Saunderson 1740 or Diderot [1749] 1961.

able to see and what he would see with the aid of visual experience (though not aided by touch). Then he laid down, stage by stage, how a patient should be prepared for a cataract operation and what kind of observations should be carried out if the experiment was to have any explanatory force. What, in fact, Diderot did was to formulate a number of methodological criteria which a psychological experiment had to meet. Finally he pointed out that patients from different backgrounds would not all react in the same manner and that perception was perhaps not as universal as Berkeley had suggested.

Like his French colleagues La Mettrie, Condillac and Diderot, the Swiss philosopher Jean-Bernard Mérian (1723–1807) wrote about Molyneux's problem, criticised Cheselden's report and suggested alternatives. All of this was contained in a series of eight *Mémoires* published between 1772 and 1782 in the *Nouveaux mémoires de l'Académie Royale des Sciences et Belles-Lettres* of Berlin.[1] In his first *Mémoire* Mérian underlined the importance of Molyneux's problem:

> Ce problème tient, dans la Philosophie moderne, une place distinguée. Les Locke, les Leibniz, les hommes les plus célèbres de notre siècle en ont fait l'objet de leurs recherches. Il a été le germe de découvertes importantes, qui ont produit des changements considérables dans la science de l'Esprit humain, et surtout dans la Théorie des Sensations.[2]

Mérian's aim was to provide an *histoire raisonnée* of Molyneux's problem in which the various solutions, including the arguments advanced and the consequences drawn, would be analyzed. In order to achieve this he posed a number of questions designed to make it possible to classify the various solutions. Are forms the direct objects of sight? Can the eye perceive shape and extension without experience? Does the eye need aid from touch in order to perceive shape and extension? Are the visual and the tactile form of an object the same form? Is similarity directly perceivable or does it have to be derived from intermediate terms with the aid of the understanding?[3]

Mérian was of the opinion that on the grounds of the solutions which he analyzed and commented upon everyone was free to choose what he regarded as appearing to be in the best agreement with the nature of things and the truth. He believed that people would answer in the affirmative if convinced that sight and touch give us the same immediate perceptions of the shape of an object or if it is possible to abstract the same ideas. However, people who believe that visual and tactile shapes are heterogeneous will give a negative

[1] In 1748 Mérian went to Berlin at Maupertuis's invitation to take a seat in the *Académie de Berlin*. In 1797 he succeeded Formey as *secrétaire perpétuel* of the *Académie*.

[2] Mérian [1772] 1984, p. 5.

[3] Mérian [1772] 1984, §2, pp. 5–8.

answer. Or perhaps they will take no sides if neither the pros nor cons seem to be decisive.[1]

We will not go into Mérian's historical review here but simply record his preference for Berkeley's solution. Mérian defended Berkeley's position that the objects of sight and those of touch are different and can only be related by experience. The fact that the words "extension" and "shape" can be applied to both visual and tactile qualities says nothing about their real identity: "nous concevrons que leur identité nominale ne prouve point leur identité réelle."[2] Like Berkeley, Mérian illustrated the nature of the link between visual and tactile objects using a language metaphor:

> Cette liaison est purement symbolique. Elle est la même qu'entre les mots et les choses, ou entre les mots écrits et les sons articulés. Les objets visibles et tangibles ne se ressemblent pas davantage que les sons ne ressemblent aux pensées, ou les mots écrits aux sons. Mais la liaison une fois établie, la présence des uns réveille l'idée des autres; comme dans nos langues les sons articulés réveillent l'image des choses signifiées, ou les caractères écrits l'image des sons qui leur sont attachés, et réciproquement.[3]

Mérian was of the opinion that the objects of sight and those of touch could fulfil the function both of sign and of the thing referred to: "De sorte que de leur nature, ils sont également propres à être, les uns à l'égard des autres, ou le signe, ou la chose signifiée."[4] This relationship is therefore mutual, said Mérian, and what first touches us always seems to summon up the idea of the other. And yet the visual object is usually given the function of sign, in view of the fact that our sense of touch is more important for our survival: "Il nous importe donc bien davantage d'être avertis par la Vue de l'effet que feront sur nous les objets tangibles, que de l'être par le Tact de l'effet que feront sur nous les objets visibles."[5] Like Locke and Berkeley, Mérian thus made the objects of sight subordinate to the objects of touch.

In order to clarify Molyneux's problem Mérian invented a new thought experiment with some resemblance to that proposed by Hutcheson. Imagine a world, said Mérian, inhabited exclusively by people born blind and in which objects give off odours more or less complex according to the complexity of their form. Would someone born without the sense of smell who suddenly has

[1] Mérian [1781] 1984, pp. 165–166.
[2] Mérian [1774] 1984, §3, p. 64.
[3] Mérian [1776] 1984, §6, pp. 93–94.
[4] Mérian [1776] 1984, §7, p. 95.
[5] Mérian [1776] 1984, §7, p. 96.

this sense restored then be able to distinguish a sphere from a cube without touching the objects? Mérian believed not.[1]

According to Mérian, Berkeley's theory had benefitted from a rare advantage that few philosophical writings can boast of, namely that of being confirmed by experience, that is by the observations of Cheselden's patient.[2] Although the young man appeared to find no similarity between visual and tactile extension and shape, Mérian wanted to cover himself against hasty conclusions: "je me garderai bien de prononcer péremptoirement d'après une expérience aussi délicate, et dont nous ne connaissons pas même les détails autant qu'il serait à souhaiter."[3]

It would be desirable, stated Mérian, if philosophers were able to perform many more authentic observations on people born blind after being operated upon, using measures laid down prior to the observations. This would enable them to compare experiences and find out from the one what the other had missed. The problem was, however, according to Mérian, that such opportunities are rare; you have to wait for a stroke of luck to come along. And not all patients are equally suitable; there is often a lack of time to prepare them and circumstances sometimes prove adverse. In addition, true researchers are no less unusual: "Les philosophes sont trop indolents, trop peu curieux pour rechercher les moyens de s'instruire; ils aiment mieux argumenter que voir. Pour le grand nombre la philosophie est un métier plutôt qu'une science."[4] In view of this state of affairs it is thus no wonder, proposed Mérian, that since Cheselden's time—forty years previous—there had been no new experiences of such an important matter.

Mérian believed that there was one major objection in the case of Cheselden's patient, as there is in that of all those born blind who undergo a cataract operation: *cataract does not cause general blindness.* And in this he was right, for even those with a mature cataract have their projection of light intact. Those blinded by cataract combine the weak light that they perceive with tactile extension, said Mérian, and thus they can have no purely visual perceptions once they have undergone the operation.[5]

Mérian suggested a bizarre plan to counteract these objections. Instead of waiting until there was a suitable case of cataract, he suggested that a

[1] Mérian [1777] 1984, §5, pp. 119–121, and Mérian [1779] 1984, §2, p. 131.

[2] Mérian [1774] 1984, §1, p. 60, and Mérian [1776] 1984, §4, p. 89. Mérian referred to Condillac's versions of Cheselden's report; he also mentioned *The Tatler* ([Steele] [1709] 1898), and contradictory observations of a person born blind operated on by Marchan. See Mérian [1782] 1984, pp. 175–177. I have only been able to check on Marchan 1768, but he says nothing about an experiment with a sphere and a cube, as stated by Mérian.

[3] Mérian [1773] 1984, part II, §7, p. 54.

[4] Mérian [1782] 1984, p. 178.

[5] Mérian [1782] 1984, p. 179.

Séminaire d'Aveugles Artificiels should be established where it would be possible to experiment systematically on a large group of people artificially deprived of the faculty of sight.[1] "Ce projet serait de prendre des enfants au berceau, et de les élever dans de profondes ténèbres jusqu'à l'âge de raison."[2] Some artificially blind people would be left to the whims of nature while others would be given a more or less good upbringing and others still would receive the best schooling that their condition would allow. They would be taught to read raised letters and would receive classes in the sciences, physics, philosophy, geometry and—especially—optics. Some would be brought up in isolation, others would be allowed to form groups. After they had attained a certain age, the cleverest would be chosen to form a society of learned men:

> En un mot, comme leur esprit serait, pour ainsi dire, entre nos mains, que nous pourrions le pétrir comme une cire molle, et y développer les connaissances dans telle succession qu'il nous plairait, on serait à portée de prendre toutes les précautions, et de varier les expériences de toutes les façons imaginables.[3]

Once they have obtained various ideas in this manner it will be easier to interrogate them one by one once the veil is lifted from their eyes, wrote Mérian.

Mérian foresaw two kinds of problem, the one physical and the other moral. In the first place one could fear that these children could lose their powers of sight completely and would really become blind. If this should turn out to be the case, the whole enterprise would, stated Mérian, be useless and would even be improper. But he could think of no fact that would confirm this possibility and he left it up to those whose task it is to study the human body.[4] Another objection could be that light is present everywhere and that no darkness is so intense that it cannot illuminate the eye when the eye has grown accustomed. Mérian brushed this problem aside by pointing out that experts would doubtless be able to invent a suitable blindfold.

The gravest objection would be that nobody wishes to sacrifice their children for metaphysical experiments, that it is unjust to require this of anyone and that it witnesses to cruelty towards those children. Mérian replied: "Je réponds que mon projet s'adresse aux philosophes embrasés de l'amour de la

[1] One could state with some measure of irony that in this Mérian was anticipating the *Institution Nationale des Jeunes Aveugles* established in 1784 by Valentin Haüy.

[2] Mérian [1782] 1984, p. 180.

[3] Mérian [1782] 1984, pp. 180–181.

[4] At present we know that visual stimulation is of essential importance in the first months following birth. Neonates brought up in darkness show serious visual defects.

Science, qui savent qu'on ne va au grand qu'en foulant aux pieds les préjugés populaires."[1]

Moreover the implementation of his plan would not be Mérian's own affair nor that of any private individual, but that of a sovereign or of a magistrate endowed with public authority. Without even speaking of the source offered by orphanages for subjects, we should allow our gaze to wander through the streets of a large town, said Mérian. Observe all those objects of our disgust and pity, these half naked children, scarcely clothed in miserable rags, the prey of vermin, brought up in laziness and debauchery, these accursed children, crippled, often even made incurably blind by monsters with pretensions to being their parents. Where is the evil in tearing these tender victims from their persecutors so that they may serve in our experiments and may be turned into useful or even famous citizens one day?

Mérian assumed that if our senses could have been developed one by one we would have derived a great deal more benefit from them. They damage each other by their simultaneous activity, since the one perfects itself at the expense of the other. Mérian would like to have control over the first impressions, since these are the most important for the general development of the individual:

> Tout dépend de ces premières séries d'impressions sensibles, et le secret de l'éducation ne roule que sur cela. Ce sont les éléments de notre raison et de toutes nos connaissances. A cela tiennent notre tour d'esprit, nos mœurs, nos penchants, notre conduite, et en grand partie notre destinée dans ce monde.[2]

According to Mérian, an individual is a bad person throughout his whole life, a suspect or a sower of unrest, only because his first impressions were bad. If our plan could be brought to fruition, said Mérian, we would be able to raise people as perfect as allowed by our nature.

The learned people who would head Mérian's *Séminaire d'Aveugles Artificiels* would have to offer their pupils objects in the order most appropriate for the end to which they were being educated. They would be able to follow the development of their understanding and in hundreds of ways develop their ideas by varying their sensations. They would be capable of making engineers, sculptors, physicists and mathematicians of the best kind, and above all philosophers unencumbered by all kinds of prejudice. In short, Mérian believed that for children raised in this way there would be infinitely more to gain than to lose. They would have no idea of what they were missing and their loss would be richly compensated:

---

[1] Mérian [1782] 1984, p. 183.
[2] Mérian [1782] 1984, p. 185.

> De quel torrent de délices vont-ils être inondés, quels seront leurs transports, lorsqu'on les fera passer de la nuit au jour, des ténèbres à la lumière, lorsqu'un nouvel univers, un monde tout brillant, éclora pour eux comme du sein du Chaos! Y-a-t-il rien de comparable à un pareil instant?[1]

In addition to all of this, science too would gain. What new discoveries could be made when children started to use their eyes, what discoveries through the association with the other senses and especially that of touch! What progress could be made in geometry, optics, natural history and philosophy! It would become possible to have them clarify the most complex questions and unravel the most aggravating problems. They would make discoveries for us by discovering themselves and it would be seen that in these places of darkness an excellent philosophical academy had been created.[2]

Would they blame you for having deprived them of their sight, Mérian asked rhetorically. Of course not: "ils vous en remercieront avec des larmes de joie, [...] ils se féliciteront, toute leur vie, de l'éducation que vous leur avez donnée."[3] And that was precisely the reaction shown by the young man who was restored to sight by Cheselden: he too greeted his benefactor with tears of joy.

Mérian's enlightened notions of the possibility of creating good, educated people seem not to have found much echo, no matter how interesting they may have been from a variety of angles.[4]

6 Conclusions

Anyone who studied the problem proposed by Molyneux was, of course, curious to know what a person born blind really would see when restored to sight. It is no surprise, therefore, that many threw themselves into discussion of Cheselden's report with great enthusiasm. Some philosophers judged that the report left nothing to be desired as far as clarity was concerned. It confirmed their suspicions that someone cured of congenital blindness would at first be unable to distinguish objects one from the other, would be unable to determine shape, size or distance, and would have to *learn* to see. The report proved, they believed, without a shadow of doubt that the man born

[1] Mérian [1782] 1984, p. 187.
[2] Mérian [1782] 1984, p. 190.
[3] Mérian [1782] 1984, pp. 190–191.
[4] According to Dilthey [1901] 1927, p. 149, Mérian's discussion of Molyneux's problem was "vielbesprochen." I was, however, unable to find any trace of this.

blind in Molyneux's question would be unable to distinguish a sphere from a cube and would be incapable of naming them. These were mostly followers of Berkeley's theory of vision, and they were largely to be found in Great Britain.

But there were other philosophers who did not allow themselves to be convinced by Cheselden's report without asking questions. They believed that the report could not be interpreted unambiguously and that the data were capable of explanation in another way. First of all they indicated the possibility that the eye was not functioning correctly, either because it was not working flexibly enough because it had not been used for a long time, or because it had not yet sufficiently recovered from the operation. Secondly they did not exclude the possibility that Cheselden had put leading questions to the boy. They accused those who interpreted the report in favour of the view held by Locke and Berkeley of reading in the report only that which was favourable to their case. Diderot pointed out that the results of the experiments were dependent on the intelligence of the patient.

Although in the seventeenth century it was known that optical correction was needed after a cataract operation, no-one blamed the results of Cheselden's operation on the absence of the lens. (However, if the sphere and cube shown to the young man were not too small and the distance not too great, dioptric factors cannot have been of any great consequence.) It is remarkable that nobody suggested that the young man could not perceive depth at first because he could only see out of one eye. His second eye was, in fact, only operated on a year later.[1]

Those who criticised Cheselden's report put forward several suggestions for getting round these objections. First of all the patient should be appropriately prepared for the operation and for the subsequent questioning. Furthermore the patient's eyes should also be given time to recover after the operation and the patient should be given the opportunity to exercise his eye muscles in darkness. Then precautions should be taken not to put leading questions. In brief, the critics of Cheselden's report—mainly to be found in France, where the influence of Berkeley's theory of vision was not so great—paid attention to a wide variety of methodological questions.

There was also a more fundamental criticism of the type of operation performed by Cheselden. Diderot believed that it made no sense at all to put questions to congenitally blind persons regarding what they see if the questioner wishes to find out how our powers of sight work. Molyneux, Berkeley and other philosophers had invented various thought experiments in the hope

[1] In all congenital cataract patients (whether operated on one eye or both) there is a lack of stereoscopic vision.

that these would provide us with insight into the working of our senses. They expected that we would be able to shrug off our prejudices and habits if we could place ourselves in the shoes of a person born blind (or of a living eye, of a statue or some such). We would then be more easily able to imagine what we would perceive before having formed our prejudices and habits. Diderot believed, however, that this was an illusory quest. He had no confidence in the reactions of someone who opened his eyes for the first time. He would have preferred to have the problems facing the theory of vision solved by a philosopher endowed with common sense.

Mérian too doubted whether cataract patients could be useful in the search for a description of pure visual perceptions. Cataract patients are, after all, not completely blind; they always retain some projection of light. However, this was not necessarily a problem within the context of Molyneux's question since this concerned only perception of shapes. Since cataract patients are not capable of perceiving shapes, they would in principle be of use in finding out whether, immediately following the operation to remove the cataracts, they could distinguish different shapes.

Mérian developed an interesting plan which would enable the (visual) development of people to be guided along pre-determined lines. He suggested having newly-born children grow up under a variety of conditions in total darkness. Once they had reached the age of reason they would be exposed to various series of visual impressions. They would then be able to provide an answer to the urgent questions regarding our powers of sight. It would, moreover, be possible to provide them with that visual information most appropriate to the end to which they were being educated. Mérian's plan for a *Séminaire des Aveugles Artificiels* was, in fact, a precursor of the deprivation experiments performed on chimpanzees after the Second World War. As will be shown in the sixth chapter, the results of the experiments were less rosy than Mérian had prophesied.

Although Cheselden's report was often referred to again in the nineteenth-century discussions of Molyneux's problem, its influence had been on the wane since the end of the eighteenth century. Admirers of the report had meanwhile been able to single out the remarkable results, while its critics had indicated the weak points.

Without there really being any question of a cæsura, around 1800 a number of developments took place which justify the suggestion that a new period was dawning in the history of Molyneux's problem. First of all there were new reports of cataract operations which were associated with the problem. Some of these provided results that agreed with Cheselden's observations while others appeared to be in conflict with them. In addition data on the visual powers of newly-born animals and babies began to be applied in discussions on Molyneux's question. Furthermore Wheatstone's clarification of stereoscopic vision came to play its part in the matter. As will be shown in the next chapter, the data mentioned above acted as arguments in the nineteenth-century empiricism-nativism debate in which Molyneux's problem frequently figured.

Chapter Five

EMPIRICAL APPROACHES IN THE NINETEENTH CENTURY

1 Congenitally Blind People Cured by Surgery

FOR several decades Cheselden's report had occupied an important place in the debate surrounding Molyneux's problem because it was—apart from Grant's dubious story—the only source of information on what a blind person could see after a successful operation. From the second half of the eighteenth century there was a slow change in this situation because other eye surgeons also began to describe the results they had achieved in operating on cataract patients.[1] Cheselden's report did not, however, fall into oblivion. Practically every time Molyneux's problem (or, more generally, the perception of people blind from birth cured of their affliction) was discussed reference was made to Cheselden. Cheselden's findings were, moreover, used as an argument in support of Berkeley's theory well into the nineteenth century.

However, in the nineteenth century Cheselden's report was also the subject of frequent negative criticism. William Hamilton (1788–1856), for instance, wrote that Cheselden's report "does not merit all the eulogia that have been lavished on it. It is at once imperfect and indistinct."[2] Samuel Bailey (1791–1870), a fervent opponent of Berkeley's doctrine, had a similar opinion: "The narrative of Cheselden, which has been so celebrated and thought to be so conclusive, appears to me, I confess, exceedingly loose, meagre, and unsatisfactory."[3] The criticism levelled at the report focused particularly on the ambiguity of certain passages. Bailey assumed that Cheselden "had not in his mind any clear idea of the import of his words, and was unaware of the ambiguity lurking in them."[4] Many found his expression "all Objects whatever touch'd his Eyes" a problem, one of the few sentences that Cheselden had reported *verbatim*.[5] Some interpreted the sentence literally and regarded it as a

[1] See Bourdon 1902, pp. 362–391, and Von Senden [1932] 1960.
[2] Hamilton in Reid [1837] 1863, vol. 1, p. 136, note.
[3] Bailey 1842, pp. 169–170.
[4] Bailey 1842, pp. 179–180.
[5] Cheselden 1728, p. 448.

proof that distance cannot be immediately seen. Others thought it extremely unlikely that Cheselden had understood the boy's words literally and thus sought an alternative explanation. Dugald Stewart (1753–1828), for example, regarded the phrase as a description of the feeling of pain caused either by the operation or by bright light.[1] Bailey supported this explanation by referring to work of the famous physiologist Charles Bell.[2] Stratton too (who, at the end of the nineteenth century, carried out ingenious experiments in which a cluster of lenses projected the outside world onto the retina the right way up rather than upside-down) was of the opinion that the patient's experience "was largely made up of muscular and tactual and pain sensations."[3]

Another explanation given of the expression in question referred to the patient's use of language. Bailey suggested that the young man was only able to use a "tactual phraseology" since throughout the whole of his life he had been familiar with objects and distances only through the sense of touch: "In seeking to express his novel feelings by the aid of an analogical language (the only resource he would have), this might be the analogy which his new sensations would most readily suggest."[4] Thomas Abbott (1829–1913) too, who also strongly opposed Berkeley's doctrine on the faculty of sight, was of the opinion that the patient's language must have been based on the sense of touch because "touch was the only mode of perception with which he was acquainted, or which he could conceive."[5] Similarly attempts were made to clarify other ambiguous phrases in Cheselden's report.

While Cheselden had merely noted down what his patient had perceived under more or less natural conditions, later doctors conducted *experiments* to verify whether, immediately following the operation, their patients could perceive shape, size, distance and suchlike. A number of doctors were particularly interested in Molyneux's question and carried out tests with a sphere and a cube to investigate how the question should be answered. The results of the cataract operations in question also drew a great deal of attention for other reasons. They were used to test Berkeley's theory of vision and served as arguments in the empiricism-nativism debate.

---

[1] Stewart [1815] 1854, p. 310, note.

[2] Charles Bell 1833, quoted by Bailey 1842, p. 173: "The beauty and perfection of the system [...] is, that each nerve is made susceptible to its peculiar impression only. The nerve of the skin is alone capable of giving the sense of contact, as the nerve of vision is confined to its own office. [...] It is most beneficently provided that this nerve (of vision) shall not be sensible to pain, nor be capable of conveying to the mind any impressions but those which operate according to its proper function, producing light and colour. The pain experienced in the eye, as from irritation of dust, is owing to a distinct nerve from that which bestows vision."

[3] Stratton 1899, p. 501, note 1.

[4] Bailey 1842, p. 174.

[5] Abbott 1864, p. 149. See also Janet 1879 and James 1890, vol. 2, p. 40.

From the numerous reports that have come down to us I will select a few points which can be regarded as important in solving Molyneux's problem. I will deal with only one case extensively, that of the Leipzig doctor J. C. August Franz, dated 1841. This is the most interesting and most accurately documented and, moreover, involved the first experiment carried out with a sphere and cube.

When we examine the various cases we note that they differ in a number of ways. First of all, the patients were not all equally blind before the operation. Some could merely distinguish between light and dark, a number were also capable of distinguishing between various colours and one or two could also perceive size, distance and motion. In a few cases there was some doubt as to whether the patient really had been born blind, or had perhaps lost the power of sight after a year or two. And some doctors even failed to report on the pre-operative state of their patients.

There was also wide divergence in the patients' age and intelligence and their emotional response varied greatly. A twenty-two-year-old female patient treated by the French doctor Jean Janin (1731–1799), for instance, gazed around her with an expression of surprise and satisfaction while saying repeatedly: "Ha, mon Dieu, que cela est beau!"[1] The forty-six-year-old female patient of the London eye doctor James Wardrop (1782–1869) was in stark contrast, appearing "bewildered from not being able to combine the knowledge acquired by the senses of touch and sight, and felt disappointed in not having the power of distinguishing at once by her eye, objects which she could so readily distinguish from one another by feeling them."[2]

Another difference was that the various doctors failed to perform their experiments at the same time subsequent to the operation and that the experiments were of different types. Jacques Daviel, whom we have already encountered as the inventor of cataract extraction, was one of the first doctors to perform experiments with his cataract patients. None of them turned out to be able to distinguish by sight the round and triangular objects which Daviel showed them.[3] Nor was Janin's patient able to distinguish objects shown to her one from the other, which Janin took to be a proof of the hypothesis advanced by Molyneux, Locke and Berkeley.[4]

Some remarkable results were obtained in tests performed by the surgeon Everard Home (1756–1832), which he described in "An Account of two Chil-

[1] Janin 1772, p. 215.

[2] Wardrop 1826, p. 533.

[3] Daviel 1762, pp. 249–250.

[4] Janin 1772, pp. 218–219. Janin—made aware of the discussion surrounding Molyneux's problem by the works of Voltaire, Diderot and Condillac—did not show his patient a sphere and a cube.

dren born with Cataracts in their Eyes" (1807). Ten minutes after Home had removed the cataract from the left eye of his seven-year-old patient he showed him a number of cards an inch in diameter and having various shapes and colours. The boy was able to name the colours correctly but he described as round not only the round cards but also the square and triangular ones. He gave a negative answer to the question of whether the cards were touching his eye but he was unable to say how far away they were. Two hours after the operation he was again shown a square and was asked if he could find corners. He had some trouble discovering a corner and then he counted the four corners of the square. In the same way he counted the triangle's corners: "in doing so his eye went along the edge from corner to corner, naming them as he went along."[1] A fortnight later the boy was still unable to distinguish shape without counting the corners one by one, "running his eye quickly along the outline, so that it was evident he was still learning, just as a child learns to read."[2] A week later the boy was able to determine the shape of the cards almost as quickly as their colours.

The experiences of Home's patient were generally regarded as a proof of intuitive seeing of distance and angles and thus as unfavourable to Berkeley's theory of vision. Bailey was of the opinion that Home's experiment had finally solved Molyneux's problem:

> It is obvious that the single circumstance of a boy, almost immediately after receiving his sight, being able to recognise a corner without the aid of touch, is a conclusive solution of Molyneux's problem in the affirmative, and it is remarkable that the proposer of it himself seems to have placed the determination of the question on this very point, alleging as a reason for answering it in the negative, that the blind man could have had no experience how a tangible angle would look.[3]

By far the most interesting case is that of Franz. This doctor was acquainted with Molyneux's problem and he was also familiar with a number of earlier reports of cataract operations. He was the first (no less than a century and a half after Molyneux had laid his problem before Locke!) actually to carry out experiments with a sphere and a cube. Franz described the results of his experiments in his "Memoir of the Case of a Gentleman Born Blind, and Successfully Operated Upon in the 18th Year of his Age, with Physiological Observations and Experiments" (1841).

Franz's patient was born squinting and with double cataracts. At the end of his second year his right eye was operated on, leaving him with an

[1] Home 1807, p. 89.
[2] Home 1807, p. 90.
[3] Bailey 1842, pp. 221–222.

atrophied eyeball. In subsequent years his left eye was operated on twice, both times without success. At the age of seventeen, when he met Franz, he was unable to perceive any light with his right eye. He was able to see light and colours with his left eye but not objects. The boy himself thought that he could partially see brightly lit objects but Franz was convinced that this was completely imaginary. He believed that the young man's behaviour was in complete accord with that of other educated blind people not completely amaurotic: they try to convince others that they see more than they really can in order to cover up their handicap and to avoid being treated with pity.

Franz removed the cataract from the left eye. Following the operation it became apparent that light caused too much pain for experiments to be performed and therefore the eye was bandaged. The bandages were taken off for the first time three days after the operation. In answer to Franz's question as to what he could see the boy said he saw "an extensive field of light, in which everything appeared dull, confused and in motion."[1] He could not distinguish objects. The pain caused by the light obliged the young man to close his eye again. When he opened it again two days later he described what he saw as "a number of opake watery spheres, which moved with the movements of the eye, but, when the eye was at rest, remained stationary, and then partially covered each other."[2] Again after an interval of two days the same phenomena were perceived, but now the spheres were less transparent and the movements steadier. The boy stated that he was able for the first time "to look through the spheres, and to perceive a difference, but merely a difference, in the surrounding objects."[3]

As soon as the young man could stand the light, Franz conducted a number of experiments. In the first series he tested whether the young man could recognise coloured ribands fixed to a black background. This proved to be the case. In the second series Franz showed the boy a piece of paper on which two strong black lines had been drawn, the one horizontal and the other vertical. After looking closely the young man was able to name the lines correctly. When he was asked to point with his finger to the horizontal line "he moved his hand slowly, as if feeling, and pointed to the vertical, but after a short time, observing his error, he corrected himself."[4] The contours of a square, in which a circle containing a triangle was drawn, was also recognised and accurately described by the boy. However, he had no notion at all of wavy or zig-zag lines.

---

[1] Franz 1841, p. 63.
[2] Franz 1841, p. 63.
[3] Franz 1841, p. 64.
[4] Franz 1841, p. 64.

In the third series of experiments, inspired by Molyneux's problem, Franz placed a sphere and a cube, each with a diameter of ten centimetres, on a level with the eye and one metre away from the patient. He then allowed the young man to move his head laterally in order to compensate for the angle of vision of the amaurotic right eye. Once he had examined the objects attentively, the patient said that he could see a quadrangular and a circular shape, and after some consideration he named the one a square and the other a disc. While the young man kept his eyes closed, Franz substituted a disc of equal size for the square and placed it next to the sphere. When the young man opened his eyes again he noted no difference in the shapes; he saw them both as discs. Franz then placed the cube in a somewhat oblique position in front of the eye and, close beside it, a figure cut out of pasteboard representing a plane outline prospect of the cube when in this position. The boy was of the opinion that both objects were something like flat quadrates. He saw as a plain triangle a pyramid that had one of its sides facing him. Franz then turned the pyramid a little so that a small area of one of its sides was visible and a large area of another. The boy looked at it for a long time and said that it was an extraordinary shape: neither a triangle nor a square nor a circle. He had no idea of it and could not describe it: "In fact," he said, "I must give it up."[1]

When this series of experiments came to an end Franz asked the boy to give a description of the sensations that the objects had produced. The youngster replied that as soon as he had opened his eyes he had discovered a difference between the two objects (sphere and cube) and had noted that they were not drawings. But he had not been able to form from them the idea of a square and a disc "until he perceived a sensation of what he saw in the points of his fingers, as if he really touched the objects."[2] When Franz placed the sphere, the cube and the pyramid in the young man's hands he was very surprised not to have recognised them as such by sight even though familiar with these mathematical shapes through the sense of touch.

Franz concluded from his experiments that objects which in reality were a good distance from the patient seemed in the beginning to be so close that he was sometimes afraid of coming into contact with them. He also believed that the young man saw everything as completely flat and far larger than he had expected. Franz also commented that his patient required spectacles in order to improve his vision.[3]

---

[1] Franz 1841, p. 65.

[2] Franz 1841, p. 65.

[3] Together with Ware, Franz was one of the few to see that cataract operations do not lead to perfect vision and that cataract glasses were needed to bring about the required correction. See Ware 1801, pp. 384–385, note.

According to Franz his experiments demonstrated that the hypothesis he had proposed elsewhere concerning Molyneux's question was correct. In his book *The Eye* (1839) he had stated that the question put by Molyneux could, strictly speaking, be answered neither positively nor negatively:

> The supposed person will certainly be able to distinguish by his sight the cube from the sphere, though he will not, it is true, recognise the two figures as a *cube* and a *sphere*, but will pronounce the one to be a *disk* and the other a *square*; it must be premised, however, that some little time must be allowed for the mind to recover from the confused sensation produced by the novelty and multitude of objects suddenly presented to the newly acquired faculty.[1]

This was exactly what Franz had noted in his patient. He underlined the importance of his study: "This is the only case on record within my knowledge wherein, with a person born blind and afterwards successfully operated upon at a period of life as far advanced as in this instance, such experiments have ever been made."[2]

Most of the reactions to Franz's report were simply congratulatory. A. W. Volkmann, for instance, wrote: "Den interessantesten Fall der Art hat Dr. Franz beschrieben [...] Diese Thatsachen sind äußerst wichtig und dürften manche Streitfragen entscheiden, über welche sich die Physiologen bis auf die letzten Zeiten nicht vereinigen konnten."[3] Abbott too praised Franz's research: "It is remarkable as the case in which the previous blindness was the most perfect, the patient the best instructed, and the observations the most accurate."[4] He added sarcastically: "yet as far as British metaphysicians are concerned it might as well have been buried in the pages of the *Illuminated Doctor*. It is ignored by one and all of them, and is only noticed (and that very imperfectly) by German philosophers; while Cheselden's inconclusive case has been sedulously copied by one from the other for more than a hundred years."[5] But Abbott voiced criticism too. He was of the opinion that the fact that the patient could not perceive a sphere and a cube as such was the fault of the experimental set-up: "Even if the young man's eye had been perfect, then, the

[1] Franz 1839, pp. 32–33, note.

[2] Franz 1841, p. 68. He added: "In the well-known case of Cheselden [...] no series of systematic experiments was instituted. Beer [Beer 1813] has also made some interesting observations, which, however, like those made in a rather superficial manner by Janin and Daviel, tend principally to describe the impressions which the newly-acquired sense had made on the mind of the person operated upon. In Ware's case the patient was not born blind, but had become so at an early period of life."

[3] Volkmann 1846, p. 268.

[4] Abbott 1864, p. 156.

[5] Abbott 1864, p. 156.

bodies were too small and too distant for a fair experiment on the perception of solidity."[1] Other criticisms related to his use of language. W. H. S. Monck blamed Franz for not having tried to investigate what his patient had meant by extraordinary expressions such as "watery spheres": "Dr. Franz's observations, though usually regarded as among the best which we possess on the subject, present features which are to me utterly inexplicable."[2]

After Franz's experiments, other doctors in their turn performed tests with a sphere and a cube. In his book entitled *On the Organs of Vision* (1858) the London eye doctor Thomas Nunneley (1809–1870) described the reaction of a nine-year-old patient. The boy was able to distinguish a difference in shape immediately, but he could in no way say which was which:

> it was only after several days that he could or would tell by the eyes alone, which was the sphere and which the cube; when asked, he always, before answering, wished to take both into his hands; even when this was allowed, when immediately afterwards the objects were placed before the eyes, he was not certain of the figure.[3]

The fact that Franz's patient had been able to name the perspective projections of the sphere and cube and Nunneley's patient had not was, according to some, because Nunneley's patient was less developed than Franz's.[4]

In 1875 the eye surgeon Arthur von Hippel (1841–1916) published an article in which he described how his patient, a four-year-old girl, reacted to a sphere and a cube:

> Sie vermochte die Körper weder richtig zu benennen, noch irgendwie ihre Form zu beschreiben; es blieb mir sogar sehr zweifelhaft, ob sie dieselben überhaupt nur als verschieden von einander erkannte.[5]

Von Hippel stated that this confirmed the correctness of the hypothesis advanced by Molyneux and Locke.[6]

Finally we should mention Ramsay's "Case of a Man Blind from Congenital Cataract who Acquired Sight after an Operation when he was 30 Years of Age" which appeared in *The Lancet* in 1903. On the second day after the operation Ramsay showed his patient a ball and a toy brick. He told him that he was showing him these objects and asked him whether he could distinguish them one from the other:

---

[1] Abbott 1864, p. 158.
[2] Monck 1882, pp. 107–108.
[3] Nunneley 1858, as quoted by Fraser in Berkeley 1871, vol. 1, p. 448.
[4] See, for instance, Stumpf 1873, p. 291.
[5] Von Hippel 1875, p. 116.
[6] See also Hirschberg 1875, p. 42.

> He looked at them attentively for a considerable time, his hands meanwhile moving nervously, as if he were trying to translate what he saw by comparing it with an imaginary tactile impression, and then he described both correctly.[1]

When asked how he had been able to distinguish the ball from the block the patient answered that "he was so much in the habit of handling objects that he had come to have a 'notion in his mind' regarding the form of things."[2] Although Ramsay did not exclude the possibility of the man having learnt on the first day after his operation what an angle was as a result of having seen a table or a chair, he suspected that the man had given the correct answer because he was capable of comparing that which he saw to that which he had felt.

Abbott regarded Ramsay's case as "the first instance in which Molyneux's question was put exactly as he suggested."[3] Indeed, Franz had not *told* his patient that he was showing him a sphere and a cube, whereas Ramsay did. And that (at least for Abbott) closed the case.

When we compare the various cases we see that we can find the entire range of possible solutions to Molyneux's problem. In a number of cases the cataract patients operated upon were incapable of seeing even the slightest difference between a sphere and a cube (or between two other objects). However, following the operation most of the people born blind were able to distinguish the objects from each other though they did not recognise them nor could they name them. One patient took the sphere for a circle and the cube for a square. Another was able to distinguish the objects one from the other and could name them once the doctor had expressly told him what he was being shown. Finally, a couple of patients were able to identify the objects correctly without further help.

It is no surprise that the reactions shown by the various cataract patients should show such a variety. The cases were, after all, very different in nature. Those wishing to make use of the experiences of the patients in order to solve Molyneux's problem therefore had at their disposal a number of cases which were contradictory and difficult to compare. The cases they used often depended on the point of view they were defending. Information agreeing with the theoretical standpoint was preferred to data that failed to fit the bill. And whenever the language used by doctor or patient was ambiguous the option was usually taken for the interpretation that fitted the point of view being defended.

---

[1] Ramsay 1903, p. 1365.
[2] Ramsay 1903, p. 1365.
[3] Abbott 1904, p. 550.

A number of philosophers were of the opinion that particularly positive case histories, that is cases where the patient was capable of identifying objects, were worthy of consideration. Thus Bailey wrote: "It is obvious that the single circumstance of a boy, almost immediately after receiving his sight, being able to recognise a corner without the aid of touch, is a conclusive solution of Molyneux's problem in the affirmative."[1] Abbott was of the opinion that "negative evidence" could have no value because it could always be blamed on unfavourable circumstances.[2] Here Abbott was referring particularly to perception of distance which, he thought, was rendered more difficult because most of the patients could only see with one eye—and this one eye did not even have a lens.[3] Another reason why the patients would be unable to identify objects, thought Abbott, could be that prior to the operation they had no clear idea of objects: "Consequently, if upon being fairly examined they appear incapable of doing so, it will follow that the defect is not in sight, but in touch; not in their new sense, but in their old ideas."[4]

Abbott also gave another explanation for why a person born blind undergoing an operation that gave him his sight would perhaps be unable to recognise objects. It is difficult enough—if not impossible—for someone with a normally developed sense of sight, argued Abbott, to form an idea of the visual appearance of an unfamiliar and complicated object relying on the sense of touch alone. A person born blind seeing for the first time would, he thought, most certainly encounter the same difficulties and would fail to recognise objects.[5] Abbott, however, considered it incorrect to conclude from this that our sense of sight can give us no idea of the actual shape and size of objects.

William James (1842–1910) also indicated problems in the assessment of the results of cataract operations. True, he was of the opinion that the solution to the problem proposed by Molyneux and Locke "has not lacked experimental confirmation. From Cheselden's case downwards, patients operated for congenital cataract have been unable to name at first the things they saw."[6] But he added: "Some of this incapacity is unquestionably due to general mental confusion at the new experience, and to the excessively unfavorable conditions

---

[1] Bailey 1842, pp. 221–222.

[2] Abbott 1864, p. 145.

[3] Even when both eyes have been operated upon there is absolutely no chance of the patient being able to perceive depth.

[4] Abbott 1864, p. 142.

[5] Before Abbott, August Zeune (1788–1853) had given a negative answer to Molyneux's question for a somewhat similar reason: "wie schwer wird es uns Sehenden, die wir das Getast noch nebenbei haben, mit zugemachten Augen eine ungewohnte Sache durchs Tasten zu erkennen, z.B. Länder auf einem Tast-Erdball" (Zeune 1821, p. 29). See also Hamilton [1858–1860] 1861–1866, vol. 2, Lecture 28, p. 176, and Mach 1886, p. 62, note 31.

[6] James 1890, vol. 2, p. 210.

for perception which an eye with its lens just extirpated affords."[1] He drew the same conclusion as Bailey and Abbott: "Obviously, positive cases are of more importance than negative."[2]

Another possibility which had to be taken into account in cases of persons born blind being operated on was pointed out in the *Essays on Philosophical Subjects* (1795) by the philosopher and political economist Adam Smith (1723–1790), published after his death. Smith supported Berkeley's theory of vision and thus started from the assumption that there is no agreement between visual and tactile objects. However, he did believe that visual objects show in some way a certain affinity with tactile objects, an affinity that we notice naturally. In this context Smith spoke of "instinctive perception." Animals, said Smith, possess a similar instinctive perception and possibly children too. But blind people operated upon would have lost this faculty:

> In him [i.e., Cheselden's patient] this instinctive power, not having been exerted at the proper season, may, from disuse, have gone gradually to decay, and at last have been completely obliterated.[3]

If Smith should be right, blind people undergoing a cataract operation would not be the ideal subjects for investigations into what we usually see thanks to nature, while this will have been one of Molyneux's intentions. Smith's finding, coupled with the problems mentioned above, constituted the reasons for other ways being taken to obtain greater knowledge of normal original visual perception. What was more obvious than to investigate the powers of sight of newborn infants or (if this should still prove too difficult and if it was regarded as justified to regard the visual perception of humans and animals as similar) why not the powers of sight of newborn animals?

2 Newborn Animals and Babies

Around the year 1800 information regarding the powers of sight of newborn animals and babies was introduced into the discussion of Molyneux's problem. This information took its place next to the wide variety of data concerning

---

[1] James 1890, vol. 2, p. 210.

[2] James 1890, vol. 2, p. 211, note.

[3] Smith 1795, p. 318. He added to this: "Or perhaps, (what seems likewise very possible,) some feeble and unobserved remains of it may have somewhat facilitated his acquisition of what he might otherwise have found it much more difficult to acquire." See also Abbott 1864, p. 163. As will be seen in the next chapter, there is what is called a critical period within which the visual system must be stimulated if seeing is to be possible.

the visual perception of persons born blind whose sight had been restored by surgical operation. The importance attached to the visual behaviour of the three groups differed from philosopher to philosopher. For instance Abbott regarded the case histories of people undergoing a cataract operation as unsatisfactory and ascribed more value to the powers of sight of babies: "[infants have] the advantage over the blind, since an instinct or faculty unemployed may tend to decay, but in the infant all the natural powers are fresh."[1] John Stuart Mill (1806–1873), on the contrary, characterised the information regarding children as "singularly inconclusive" and that on blind people operated on for cataract as "the most valuable facts of all."[2] Bailey, finally, called blind people cured of cataract "perhaps the most interesting source" and regarded certain forms of behaviour exhibited by young animals as a decisive proof against Berkeley's doctrine of sight and thus as a denial of his answer to Molyneux's question.[3]

As will be seen, there was also a difference of opinion as to whether the power of sight possessed by the cured blind, young animals and babies was similar or not. Some believed that information relative to the one group could be extrapolated to the other, while others pointed to fundamental differences between the three groups.

Researchers generally agreed that certain animals can perceive objects at various distances immediately after birth. Adam Smith illustrated this using the behaviour of chickens, partridges, grouse and suchlike which, as soon as they emerge from the egg, walk around the field calmly and pick up the seeds indicated by their mothers: "they no sooner come into the light than they seem to understand this language of Vision as well as they ever do afterwards."[4] This also applied, said Smith, to animals which open their eyes only a few days after birth and for certain beasts, such as calves and foals. Thomas Brown held a similar opinion: "The calf, and the lamb, newly dropt into the world, seem to measure forms and distances with their eyes, as distinctly, or at least almost as distinctly, as the human reasoner measures them, after all the acquisitions of his long and helpless infancy."[5]

According to the German physiologist and comparative anatomist Johannes Müller (1801–1858), the fact that some newborn animals immediately see their mother's nipples proves that the power of seeing simple shapes is not acquired. And thus he failed to understand why Molyneux and Locke had

[1] Abbott 1864, p. 163.
[2] Mill [1842] 1859, p. 106 and p. 110, respectively.
[3] See Bailey 1842, p. 149 and p. 166.
[4] Smith 1795, p. 319.
[5] Brown [1820] 1860, Lecture 28, p. 181.

answered Molyneux's question in the negative.[1] Bailey reached a similar conclusion. He stated that the movements made by certain animals immediately following birth indicate that they can see objects at various distances:

> Their running about, their snatching at objects presented to them as soon as born, their seeking the teats of the dam, their leaping from one spot to another with the greatest precision, all show not only that they can see objects to be at different distances, but that there is a natural consent of action between their limbs and their eyes, that they can proportion their muscular efforts to visible distances.[2]

In order to underline his opinion Bailey quoted Humphrey Davy and Cuvier. Davy had noted that crocodiles bite a stick held in front of them as soon as they emerge from the egg. And Cuvier had seen that in the first few days following the birth of monkeys they remain hanging from their mother's breast and look intently at all kinds of objects without touching them. When they first jumped towards or grasped something they showed that they could estimate distance exactly.[3] Bailey stated that these examples justified his reasoning:

> Here then we have positive proof that a perception of degrees of distance is immediately possessed at birth through the unassisted organs of vision—through organs constructed in all respects essential to the present argument like the human eye.[4]

The observations of animals were regarded by many philosophers as an argument against Berkeley's theory of vision since they believed that his doctrine was not especially based on sensual processes actually taking place in neonates but on the assumption that it is *impossible in principle* to perceive distance, whether an animal or a human eye is involved.[5] A single case of an animal being capable of seeing distance from birth would, they thought, constitute a good counter to this assumption.

Ironically enough it happened to be a fierce defender of Berkeley's doctrine, Adam Smith, who had advanced this argument.[6] Hamilton was one of the first to take up the argument. He concluded that

---

[1] Müller 1837–1840, vol. 2, p. 362.

[2] Bailey 1842, pp. 149–150; see also Bailey 1842, pp. 29–30.

[3] Bailey 1842, p. 150. Abbott presented similar examples taken from Smith, Bell, Cuvier and Flourens. He also believed that most animals have no special organs of touch and that their only idea of shape and size must therefore be visual. See Abbott 1864, pp. 163–173.

[4] Bailey 1842, p. 151.

[5] See, for example, Bailey 1842, p. 38 and p. 152, and Abbott 1864, p. 167. Bailey spoke in this context of "a sort of mathematico-metaphysical argument."

[6] Smith called Berkeley's theory "one of the finest examples of philosophical analysis that is to be found, either in our own, or in any other language." Smith 1795, p. 294.

> [Berkeley's] theory is provokingly—and that by the most manifest experience—found totally at fault with regard to them [i.e., animals who possess at birth the power of regulated motion].[1]

Bailey too was of the opinion that the proof obtained from animal behaviour was "complete and conclusive."[2] Abbott was of a similar opinion.[3]

Despite all this criticism, one or two individuals took great pains to hold up Berkeley's theory. John Stuart Mill, for example, agreed that it was true that the data derived from young animals "have long been felt to be a real stumbling-block in the way of the theory."[4] But he regarded them as little more than a difficulty rather than a complete denial of Berkeley's doctrine.[5] Moreover Mill found fault with the conclusion based on some types of behaviour on the part of young animals that they can perceive distance immediately after birth. Kittens and puppies, stated Mill, are born blind and yet they find their mother's teats before their eyes open.[6]

Though there may have been agreement regarding the nature of the faculty of sight in newborn animals, researchers seemed to have problems agreeing what newborn babies can see. It was not without reason that Abbott wrote that "nothing [...] is more difficult than tracing the earliest development of ideas in the mind of an infant."[7] Experimental developmental psychology had not yet got off the ground; there were scarcely any effective research methods and too little had at the time been published on babies' powers of sight. The few publications which did exist coupled with observations made by the researchers themselves did persuade some to conclude that neonates are not capable of seeing objects at various distances immediately after birth. But others believed that, like newborn animals, they were capable of such activity.

Thomas Brown (1778–1820) was one of those who assumed that man's powers of sight were substantially different from those possessed by animals.

[1] Hamilton in Reid [1837] 1863, vol. 1, p. 182, note.

[2] Bailey 1842, p. 149.

[3] Abbott 1864, p. 167. Abbott characterised Berkeley's theory as "the shame, not the glory, of psychology" (p. 2).

[4] Mill [1842] 1859, p. 107.

[5] Mill [1842] 1859, p. 110. In his response to Bailey, Mill wrote that "Berkeley's Theory of Vision, has remained, almost from its first promulgation, one of the least disputed doctrines in the most disputed and most disputable of all sciences, the Science of Man. [...] If the doctrine be false, there must be something radically wrong in the received modes of studying mental phenomena" (pp. 84–85).

[6] Mill [1842] 1859, p. 109. Mill took this argument from the *Spectator*.

[7] Abbott 1864, p. 163. Johannes Müller had written at an earlier date: "Es ist ungemein schwer, wenn nicht völlig unmöglich, sich mit einiger Wahrscheinlichkeit einzubilden, wie das Kind die ersten Eindrücke auf die Nervenhaut des Sehorgans beurtheilt." Müller 1826, quoted by Hirschberg 1875, pp. 27–28.

Brown believed that the fact that some animals can instinctively perceive distance proves that there is no physical impossibility involved in supposing that a similar "original suggestion" may take place in man. But he believed that experience teaches us that a great deal of time is required before infants are capable of distinguishing different objects by sight and to fix their gaze on things. Hence his conclusion that "in man, there is not that necessity for the instinct, which exists in the peculiar situation of the other animals; and we find, accordingly, that there is no trace of the instinct in him."[1] Brown believed that the situation of babies could be compared to that of people born blind whose sight is restored at a later age and cannot distinguish a sphere from a cube: in both cases the actual size, shape and position of objects has to be learnt in the manner of a new language. This showed, according to Brown, "that we learn to see,—and that vision is truly, what Swift has paradoxically defined it to be, the *art of seeing things that are invisible*."[2]

Other philosophers, however, believed that the human faculty of sight was indeed similar to that of animals. Thus Adam Smith found it difficult to imagine that man would be the only animal whose young would have no instinctive perception. The young of humans are, however, dependent on others for such a long period, they have to be carried in the arms of their mothers or nurses so long, said Smith, that instinctive perception of this kind would seem to be less necessary for them than for other species. Before it came to have any useful purpose, observation and experience would have sufficiently forged the connections between visual and tactile objects. But because children appear to see the distance, shape and size of objects at a very early stage, Smith tended to believe that even they enjoyed some kind of instinctive perception, "though possibly in a much weaker degree than the greater part of other animals."[3] As already mentioned, Smith attributed to decay of the powers of sight the failure of blind people operated on for cataract to immediately see objects at a distance.

Bailey too believed that babies' powers of vision were similar to those enjoyed by young animals. And while Bailey was convinced that babies did not possess full perception of distance immediately after birth, he still thought that this did not prove that of its nature the eye was unable to perceive distance, as Berkeley had stated. Failure to perceive in a complete manner would, he believed, be attributable to the *immaturity* of the organ.[4] And even with regard to his other senses and limbs, thought Bailey, man was at

---

[1] Brown [1820] 1860, p. 181.

[2] Brown [1820] 1860, p. 181. See also Stewart [1815] 1854, p. 338, Preyer 1882, pp. 39–40, Preyer 1908, p. 37, and Raehlmann 1891, pp. 54–56.

[3] Smith 1795, p. 323.

[4] See also Abbott 1864, p. 163, and Mach 1900, p. 93.

a disadvantage: it takes time for all human organs to mature. But as soon as this is achieved, the organs can perform all the functions for which they were created. Thus Bailey was quite sure that "the power of performing all the functions of sight is in the eye as soon as it has come to maturity."[1]

Physiologists and mothers, stated Bailey, had noticed that babies open their eyes immediately following birth and that they quickly start directing their gaze towards sources of light, at first by turning their heads and later also their eyes towards these sources. They had then noted that babies very quickly demonstrate interest in faces and see relative sizes and distances. The studies performed had also shown, said Bailey, that touch remains inert for a long time and develops later than sight: babies do not start using their hands until they are a few months old and it takes a long time for them to point in the right direction or to estimate distance.[2]

When babies use their hands for the first time, said Bailey, they attempt instinctively to touch what they see. In the beginning they fail, for they need to learn how to adapt their muscle power to visual distances:

> Here is no process of learning to see with precision by the help of the touch, but one of learning to touch with precision by the help of the sight. It is, in some respects, the reverse case to that of the blind man who receives his sight from a surgical operation.[3]

A person born blind and cured surgically is required to learn to connect his new visual sensations and his old tactile sensations, while babies need to do the reverse: to connect their new tactile sensations and their old visual sensations. According to Bailey the attempt made to grasp an object implies that the object is seen at a certain distance. On the basis of the above, Bailey drew the following conclusion:

> In the progress of the human infant, then, it is clear that the priority of definite perceptions of extension is with the sight, and that in the connection which is soon established between his visual and tactual sensations, the process is so far the reverse of what Berkeley's theory requires it to be.[4]

As far as he was concerned this constituted a supplementary argument in favour of a positive answer to Molyneux's question.

---

[1] Bailey 1842, p. 155.

[2] Bailey referred to the *Traité de psychologie* by C. F. Burdach (translated from German), to *Progressive Education* by Mme Necker de Saussure (translated from French) and to *The Hand* by Charles Bell. See Bailey 1842, pp. 158 ff.

[3] Bailey 1842, p. 163.

[4] Bailey 1842, p. 165.

Abbott used a process of analogous reasoning to reach the same conclusion: "There is, indeed, no instance in which we are more justified in reasoning from the lower animals than the present. Sight is the most universal special sense."[1] Abbott believed that all vertebrates can see distance immediately and that because of the uniformity shown in the construction of the eye it was therefore extremely unlikely that human beings were unable to do so:

> The greater the resemblance between the organs of sight and the general phenomena of the nervous system in men and beasts, the greater this improbability. As every advance in physiology tends to develop this resemblance and to strengthen the conception of the unity of system in the animal kingdom, this argument has also become stronger.[2]

Abbott believed that careful observation had demonstrated that "there is an independent visual perception of distance and magnitude in the earliest infancy prior to any association with touch or locomotion."[3] He quoted with approval the physiologist Schroeder van der Kolk who regarded it as absurd that children were said by certain writers to gain their first impressions of distance and size viâ the sense of touch and learnt to see through touch. He was himself convinced that children see objects at different distances long before they take hold of them with their hands and begin to explore them.

The above shows that information relative to the powers of sight of newborn animals and babies was not only advanced in support of general theories of vision but was also used as an argument for various solutions to Molyneux's problem. Some regarded the fact that newborn animals can immediately see objects at a certain distance as supporting an affirmative answer to Molyneux's question. Others, on the contrary, believed that visual behaviour on the part of babies pointed to the fact that seeing is an acquired skill and that a person born blind would not be able to identify a sphere and a cube immediately after having his sight restored by surgery. Information on the powers of sight of newborn animals and babies, therefore, was just as ineffectual in producing agreement on Molyneux's problem as had been the information on the powers of sight of persons born blind seeing after an operation.

Up to this point philosophers and other scientists, when justifying their various solutions regarding Molyneux's problem, had referred to the behaviour

---

[1] Abbott 1864, p. 167. Cf. Hamilton [1858–1860] 1861–1866, vol. 2, Lecture 28, pp. 181–182.

[2] Abbott 1864, p. 168. Abbott left open the possibility that human beings have to learn from experience what animals perceive directly, but he did not believe that anyone had demonstrated that this was the case.

[3] Abbott 1864, p. 166.

of creatures seeing for the first time, which is not all that surprising.[1] Halfway through the nineteenth century, however, a particular aspect of the perception of *experienced* observers was advanced as justification of the various solutions to the problem, namely stereoscopic perception of depth.

3 Stereoscopic Perception of Depth

In the ancient world the perception of depth was regarded as self-explanatory.[2] At the beginning of the seventeenth century a change occurred when Kepler discovered that the outside world was projected onto the retina in reverse and flat. From that time the question began to be posed as to how it was possible to obtain *depth* from a *flat* retinal image? More than two centuries later the all-round inventor and physicist Charles Wheatstone (1802–1875) succeeded in providing an answer.

Wheatstone described his discoveries in an article in the *Philosophical Transactions* which became very famous.[3] Here he showed with the aid of the stereoscope, which he had himself invented, that perception of depth is achieved by joint action of non-corresponding retinal points.[4] To demonstrate this, Wheatstone had made two line drawings of the same object: one from the angle of vision of the left eye and the other from the angle of vision of the right eye. When these two drawings were looked at through a stereoscope, there was an unmistakeable impression of a solid three-dimensional object. (Wheatstone remarked that different retinal images can be created only when the object looked at is close to the eyes. When an object is looked at from some considerable distance the optical axes of both eyes are parallel and the perspective projections are the same on both retinas.[5])

---

[1] Abbott 1864, p. 140, for instance, spoke of "cases in which the phenomena of sight are given most pure and independent—viz., those of infants, of persons born blind, who have been enabled to see, and of the lower animals." Cf. Bailey 1842, p. 148.

[2] See Crone 1989.

[3] Wheatstone 1838.

[4] In order to explain why an object looked at with both eyes is seen as single, Christiaan Huygens had defined the so-called corresponding points: "chaque point du fond de l'œil a son point correspondant dans le fond de l'autre en sorte que lors qu'un point de l'object est peint dans quelques deux de ces points correspondants, alors il ne parait que simple comme il est" (Huygens [1667] 1916). Wheatstone rejected the suggestion that seeing a single image was possible *only* with corresponding points, since objects depicted on non-corresponding points of the retina do not have to be seen as double. If they are seen as single, they create a sensation of depth.

[5] Wheatstone stated that this explained why an artist can never make a true-to-life image of a nearby object. In the case of a painting, two identical images are projected onto

Wheatstone's discovery was a heavy blow to Berkeley's theory of vision. Bailey was among those quick to realise this:

> A striking argument to this effect [i.e., that we originally see objects at various distances] may be derived, if I mistake not, from the recent discoveries of Professor Wheatstone in binocular vision. [...] Here, then, there is a certain combination of impressions on the nerves of the eyes, followed by a perception of geometrical solidity, even contrary to the testimony of the sense of touch, proving that the perception of the third dimension of space by the sight is immediate, and independent of information acquired by any other sense.[1]

Bailey also linked Wheatstone's discovery to Molyneux's problem. When a blind person suddenly acquires vision and looks at a nearby solid object, stated Bailey, two different retinal images are formed of it "and the perception of an object of three dimensions would be doubtless produced in his mind."[2]

John Stuart Mill, attempting to turn aside Bailey's attack on Berkeley, said that he thought the former was attaching too much importance to Wheatstone's findings.[3] The phenomena described by Wheatstone were, he said, consistent with both the theory of Berkeley and that of Bailey. "If either theory could derive support from this experiment," Mill added, "it would surely be that which supposes our perceptions of solidity to be inferences rapidly drawn from visual impressions confined to two dimensions."[4] Mill opted for this opinion because, he thought, we can see nothing except in so far as it is represented on our retina.[5] Wheatstone's experiment had, however, demonstrated convincingly that in the spatial perception of a three-dimensional object the condition of *both* retinas is important. It had also demonstrated that we can *see* objects spatially without there being any need to involve the sense of touch, and this did not rhyme with Berkeley's opinions on the subject. Bailey also remarked that a stereoscope can even give us an impression of relief, whereas the sense of touch merely tells us that there are two flat sheets

---

the retina. Leonardo da Vinci had earlier discovered that a painter will never succeed in perfectly reproducing the sensation of depth. In order to provide the best illusion of depth he advised looking at a picture from some distance with one eye.

[1] Bailey 1842, pp. 100–101.

[2] Bailey 1842, p. 101. This is a false assumption in view of the fact that binocular vision suffers more greatly and in a more definitive fashion as a result of visual deprivation than does monocular acuity.

[3] Mill [1843] 1859, p. 117: "Mr. Bailey, in his reply, insists very much on [...] the confirmation which he imagines his theory to derive from Mr. Wheatstone's discoveries respecting binocular vision, exhibited in the phenomena of the stereoscope." Bailey's pamphlet was published as Bailey 1843.

[4] Mill [1843] 1859, p. 118.

[5] Mill [1843] 1859, p. 116.

of paper. Although Wheatstone's explanation rapidly gained general acceptance, there was disagreement as to whether stereoscopic or spatial vision is acquired or inborn. This was the question at the heart of the nineteenth-century empiricism-nativism debate.

4 The Empiricism-Nativism Debate

In the second half of the nineteenth century physiologists and psychologists concentrated on the question of whether our spatial perception is acquired or inborn. Those known as nativists assumed that by nature we see objects in a spatial relationship because we have an inborn knowledge of the order inherent in retinal points. Empiricists disagreed; they believed that we merely learn to recognise the spatial characteristics of objects. Both empiricists (Lotze and Helmholtz being the best known) and nativists (including Müller, Panum and Hering) were inspired by Kant's doctrine of the *a priori* nature of space.

Immanuel Kant (1724–1804) had no hesitation in stating that all our knowledge begins with experience, but he did not believe that all knowledge stems from experience.[1] What he was attempting to show in his *Critik der reinen Vernunft* was that although our knowledge cannot rise above experience, it is nonetheless partially *a priori* and cannot be derived inductively from experience. Kant assumed that the *Ding an sich*, which is the cause of our sensations, is unknowable. The external world provides the material for our sensations; our capacity for knowledge arranges it in time and space and provides us with the concepts which help us to comprehend our experiences. Kant distinguished two sources of human knowledge: the *Sinnlichkeit* and the *Verstand.* Objects are *given* to us by means of our senses; but they are *thought* by our understanding. In the former case it is a question of forms in which the entire experience can be placed, namely space and time. In the latter it has to do with categories of our understanding, concepts which make possible our experiential knowledge.[2]

Space, said Kant, is not a perception in itself; it is a form of perception, a way of intuition. Space is subjective; it is part of our equipment of perception. And thus everything we perceive will bear spatial characteristics. We use our external senses to suggest to ourselves that objects are outside us and in their entirety in space. Space is not an empirical concept abstracted from the

---

[1] He quoted Cheselden's patient in a discussion of this point: "nun muß man, wie der Blinde des Cheselden fragen: was betrügt mich, das Gesicht oder Gefühl?" Kant 1788, "Vorrede," p. 27.

[2] See Kant [1781] 1787, bk. I, part 1, §i.

external world, since space is already assumed when we relate our sensations to something external. Space is a necessary representation *a priori*, which is at the basis of all external intuitions. Indeed, we cannot imagine that there should be no space. Space is not a discursive or general concept of relations between the *Dinge überhaupt*, but a pure intuition; there is only one space. Space is represented as an infinitely given magnitude. In short, according to Kant space does not represent a characteristic of one or other *Ding an sich*; it is nothing other than the form of all phenomena of the external senses, that is to say, the subjective precondition of our sensory nature.[1]

Kant's theory only referred to space as a necessary representation *a priori*; it had no connection with the question regarding the intuition of *empirical* space as inborn or acquired.[2] This was the central question in the controversy between empiricists and nativists. In order to provide solid foundations for their points of view, researchers used not only metaphysical and methodological arguments but also information on the powers of sight of newly born animals, infants and blind people operated on for cataract, and within this context Molyneux's question once more came up.[3] The empirical evidence was, as previously, subjected to a variety of interpretations. Stumpf, for example, wrote as follows: "Nebst den operirten Blindgeborenen hat man auch die Kinder und die jungen Thiere zu Zeugen aufgerufen, und daß diese, wenn überhaupt für eine, für die nativistische Theorie zeugen, versteht sich von selbst."[4] Others, including John Dewey, proclaimed the contrary: "The proofs of this theory of the acquired nature of sight perception of space are found in the observations made upon infants, and upon the congenital blind, when given sight."[5]

The solutions offered for Molyneux's problem were to some extent related to the point of view that each protagonist took in the empiricism-nativism debate. Thinkers who were inclined towards empiricism usually believed that a person born blind who suddenly recovers the powers of sight would be unable to distinguish a sphere from a cube, let alone being able to name them.[6]

---

[1]See Kant [1781] 1787, bk. I, part 1, §§ii–iii.

[2]Hamilton wrote, for instance, "The *a priori* Conception [of Space] does not exclude the *a posteriori* Perception." Hamilton in Reid [1837] 1863, vol. 1, p. 126, note.

[3]Davis 1960, called Molyneux's problem "the progenitor of the nativism-empiricism controversy."

[4]Stumpf 1873, p. 294. See also Müller 1837–1840, vol. 2, p. 362, Panum 1858, pp. 87–88, Mach 1900, p. 93, Dunan 1888, p. 382, and Ebbinghaus 1908, p. 59.

[5]Dewey [1887] 1891, p. 145. See also Hirschberg 1875, pp. 27–28, Preyer 1882, p. 39, Raehlmann 1891, pp. 54–60 and p. 95, and Wundt [1896] 1907, pp. 159–160.

[6]See, for example, Helmholtz 1856–1867, vol. 3 (1867), pp. 593–594, Hirschberg 1875, p. 24, p. 28 and p. 42; Von Hippel 1875, p. 116, Preyer 1882, p. 39, and Dewey [1887] 1891, p. 145. Raehlmann was one of the few empiricists to conclude, on the basis of his own studies of a person born blind who recovered the faculty of sight thanks to an operation,

Thinkers inclined to the nativist viewpoint came to a variety of conclusions. Some of them believed that the person born blind would be able to do no more than distinguish the objects from one another, while others assumed that he would also be able to name them.[1] A few nativists believed that the solution to Molyneux's problem had indeed been experimentally proved but they ascribed the incapacity shown by the patients involved to mental confusion and unfavourable circumstances, rather than to the lack of an inborn representation of space.[2]

The difference between empiricists and nativists was, in fact, not as great as it was made to seem. A. W. Volkmann noted this at an early stage when he wrote the following:

> Daß unsere Raumanschauungen nicht minder Anerzogenes als Angeborenes enthalten, darüber dürfte unter den Physiologen und Psychologen kaum ein Zweifel vorkommen, nur die Grenzen beider sind streitig. [... Es] scheint mir unzweifelhaft, daß die Fasern der Tast- und Sehnerven das Vermögen räumlich verschiedene Eindrücke zu erzeugen, von Geburt an besitzen. Zwar ist unverkennbar, daß wir die Raumverhältnisse der objectiven Welt erst durch Erfahrungen kennen lernen, aber dieses Erfahren setzt ein Vermögen, Erfahrungen zu machen, schon voraus. [...] Ich halte also die Wahrnehmung des Extensiven ebenso für eine primitive Thätigkeit des Sehnerven, wie die Wahrnehmung des Farbigen.[3]

Kant's notion of space was not only a source of inspiration for empiricists and nativists but also led to the question of whether those born blind do actually have any notion of space. In a certain sense this too brought one of Molyneux's assumptions into the discussion.

Friedrich Gottlob Born, an early Kantian, was one of those who started off with the assumption that those born blind and the sighted have the same

---

that such a person would be able to distinguish these two objects one from the other but would be unable to name them. See Raehlmann 1891, p. 94.

[1] See, for example, Janet 1879, p. 7, respectively Tourtual 1827, Müller 1837–1840, vol. 2, p. 362, and Stumpf 1873, p. 291. Stumpf was of the opinion that the man would be able to name the objects provided that he was educated.

[2] James 1890, vol. 2, p. 210, Mach 1886, p. 62, and Schlodtmann 1902, p. 260, note.

[3] Volkmann 1863, pp. 139–140. Hering had a similar opinion: "Zwischen 'Nativismus' und 'Empirismus' besteht kein grundsätzlicher, sondern nur ein gradweiser Unterschied. Wenn uns, um dies hier abermals abzusprechen, die Organe angeboren sind, so sind es bis zu einem gewissen Grade auch ihre Functionen, das müssen selbst die strengsten 'Empiristen' zugeben; und andererseits hat es nie einen 'Nativisten' gegeben, der den gewaltigen Einfluß geleugnet hätte, welchen Gebrauch und Übung auf die Functionen unserer Organe und besonders der Sinnesorgane hat." Hering [1874] 1878, as quoted in Schlodtmann 1902, p. 257.

representation of space to the extent that space is a pure intuition. But he believed that the representation that the blind have of partial, empirical space is totally different from that which the sighted have, since they learn to know objects exclusively by touch. On these grounds he regarded discussion of Molyneux's problem as a superfluous pursuit.[1]

On the other hand, in 1793 the German philosopher and medical man Ernst Platner (1744–1818) launched the suggestion that the blind have absolutely no notion of empirical space:

> die Beobachtung und Untersuchung eines Blindgebohrnen [hat mich] überzeugt, daß der Gefühlsinn für sich allein alles dessen, was zu [...] Raum gehört, durchaus unkundig ist, nichts von einem örtlichen Außereinanderseyn weis; [...] Wirklich dient dem Blindgebohrnen die Zeit statt des Raumes. Nähe und Entfernung heißt bey ihm weiter nichts, als die kürzere, oder längere Zeit, die geringere, oder größere Anzahl von Gefühlen, die er nöthig hat, um von einem Gefühl zum andern zu gelangen.[2]

Any temptation on our part to regard this as strange is due, said Platner, to the fact that we allow ourselves to be deceived by the *visual language* used by those born blind. Platner considered his opinion to have been proved not only by the blind people he studied but also by Cheselden's patient: if this man had had a notion of space before the operation, everything that was far away would not have seemed close nor would all the separate objects have seemed like one great whole.[3] To quote James: no opinion is so silly but it will find some "learned Theban" to defend it.[4] In 1844 F. W. Hagen expressed his agreement with Platner.[5] And Hamilton expected that if a blind person had formed the idea of a sphere and a cube by touch, he should be able to distinguish them from one another on seeing for the first time. But Cheselden's patient had not been able to do so and on that basis Hamilton concluded that only sight can provide us with empirical representations of space.[6] Abbott

---

[1] Born 1791, pp. 110–111: "Man darf daher nicht, mit Molyneux, die Frage aufwerfen: ob ein Blinder, der Kugel und Würfel durchs Gefühl gekannt hatte, sie auch durchs Ansehen kennen würde, wenn er plößlich das Gesicht bekäme? [...] Denn da das Gesicht und das Gefühl von denselbigen Gegenständen nicht auf einerley Art afficirt werden; so muß daher die empirische Raumvorstellung von körperlichen Räumen in Personen, bey welchen sie durch das einseitige Gefühl bestimmt wird, ganz anders ausfallen, als bey Leuten, die Gesicht und Gefühl verbinden."

[2] Platner [1776] 1793, §765, p. 440.

[3] Platner [1776] 1793, §765, p. 441.

[4] James 1887, p. 211.

[5] Hagen 1844, p. 718: "Was uns Raum ist, ist bei [Blindgeborenen] bloß Zeit." See also Lotze [1881] 1882.

[6] Hamilton [1858–1860] 1861–1866, vol. 2, Lecture 28, pp. 176–177.

had come to a similar conclusion.[1] John Stuart Mill also subscribed to Platner's suggestion that the idea of space could eventually be reduced to that of time and, furthermore, he believed that Platner's observations had demonstrated that the concept of extension or distance is generated by the series of sensations of muscle movements, as Bain had proposed.[2] Platner found a fierce defender even halfway through the twentieth century; the German psychologist Max von Senden thought Platner to be "the first to conclude that we merely allow ourselves to be deceived by the verbal habits of the blind and that in reality they have no awareness of space."[3]

As could be expected, there was also criticism of Platner.[4] The French philosopher Charles Dunan (1849–1931), while regarding Platner's observations as confirmation of the thesis that the blind would lack the notion of space possessed by the sighted, said that this did not mean that they would have no notion of space at all, as Platner had stated. Through discussions that he himself had had with blind people Dunan was convinced that the blind certainly do have a good notion of space; it is simply different from that of the sighted. On these grounds Dunan responded in the negative to Molyneux's question.[5]

James thought that Dunan, in his turn, had only been partially right. He believed that the notion the blind have of space may well differ from that of the sighted, but that at the same time there is a deep analogy between the two:

> "Big" and "little," "far" and "near," are similar contents of consciousness in both of us. But the *measure* of the bigness and the farness is very different in [the blind man] and in ourselves. He, for example, can have no notion of what we mean by objects appearing smaller as they move away, because he must always conceive of them as of their constant tactile size. Nor, whatever analogy the two extensions involve, should we expect that a blind man receiving sight for the first time should recognise his new-given optical objects by their familiar tactile names.[6]

---

[1] Abbott 1864, p. 151 and p. 161.

[2] Mill 1865, p. 231. See also Bain 1855.

[3] Von Senden [1932] 1960, p. 28. (In view of the fact that most of the copies of the original work were lost in the Second World War, I used the English translation of 1960.) On page 286 Von Senden wrote: "Since nothing is given simultaneously to his [i.e., the blind person's] senses as spatial, it must be mentally strung together in time, which does duty for the spatiality he lacks."

[4] Platner has, more recently, also been criticised by Evans 1985.

[5] Dunan 1888, p. 135.

[6] James 1890, vol. 2, p. 210. See also Mach 1886, p. 62, note 31, and Mach 1900, p. 100, note 2.

James was therefore not surprised that the answer given by Molyneux and Locke to Molyneux's question had been experimentally confirmed.[1]

What James was trying to make clear was that there are different concepts of space. One of the major differences between visual and tactile space is that visual space is projective and non-Euclidean whereas tactile space is Euclidean. In the case of sight it can be said, for instance, that an object reduces in size in proportion to its distance from the observer, that angles change according to the point of view and that circles can pass viâ ellipses into straight lines. This does not apply to the sense of touch: angles, lines and surfaces are unchangeable, no matter how an object is rotated.

James was convinced that the concepts of space originally different from each other and even incoherent are eventually through experience reduced to a common measure, namely that of the real world.[2] At the end of the nineteenth century Stratton demonstrated experimentally that tactile localization, as regards both direction and distance, can be influenced by a disturbance of normal visual localization. This caused him to conclude that through association a correspondence between touch and sight is formed point-by-point and that we learn only through experience which visual position corresponds to which tactile position. Stratton saw confirmation of his opinion in the reactions shown by blind people after an operation.[3]

5 Conclusions

When we take an overall view of the developments described above, we can first conclude that Molyneux's problem had a less prominent position in the nineteenth than in the eighteenth century. While philosophers such as Berkeley, Condillac and Diderot had taken the problem as the starting-point for their psychology of perception or doctrine of knowledge, nineteenth-century researchers generally treated it as an interesting question within the context of theories of spatial perception. The arguments advanced in favour of the various solutions in the nineteenth century were not so much related to the link

---

[1] James 1890, vol. 2, p. 210. As already stated, James partially ascribed the incapacity of the patients in question to unfavourable circumstances and attached more importance to positive than to negative cases. See also Villey 1914, pp. 168–184.

[2] "The various space-senses are, in the first instance, incoherent with each other; [. . .] The *education* of our space-perception consists largely of two processes—reducing the various sense-feelings to a common *measure*, and *adding them together* into the single all-including space of the real world." James 1887, pp. 536–537; also in James 1890, vol. 2, pp. 268–269.

[3] Stratton 1899. Stratton conducted an experiment in which mirrors were used to project the image of the test subject diagonally before him.

between visual and tactile sensations or between visual and tactile concepts, as had previously been the case, but were more involved with empirical data concerning the perception of blind people restored to sight, newborn animals and babies.

It is remarkable that there was still no agreement as to what was the most plausible answer to Molyneux's question. New cataract operations performed on people born blind provided results that were contradictory and difficult to assess. If we have to indicate a general tendency, we could say that most researchers believed that a person born blind receiving his sight would be able to distinguish a sphere from a cube but would not be able to name the objects correctly.

One of the principal shifts in the nineteenth century was that Berkeley's theory of vision (and therefore his solution to Molyneux's problem) lost its ability to convince. Information on the powers of sight of newborn animals (and to a lesser extent that of human infants) and Wheatstone's explanation of stereoscopic vision contributed to this. Animals were seen to be capable of perceiving objects at different distances immediately after birth, infants showed evidence of spatial vision before any tactile experience worthy of mention and Wheatstone demonstrated that stereoscopic perception of depth was caused by retinal disparity. In short, it became clear that spatial vision was not derived from touch, as Berkeley had stated.

There was disagreement surrounding the question of the extent to which spatial vision was more acquired than inborn. As we shall see in the next chapter, discoveries were about to be made in the field of neurophysiology which would provide more clarity in the matter, just as research into visual deprivation was to provide greater insight into the way to assess data derived from blind persons treated surgically.

CHAPTER SIX

MODERN APPROACHES

1 HISTORICAL ANALYSES

WE have seen that Molyneux's problem occupied a central position in eighteenth-century epistemology and psychology and that in the nineteenth century its main rôle was in theories of spatial perception and of the formation of spatial concepts. This chapter will investigate the importance that twentieth-century researchers have attached to the problem and the methods they have employed to solve it.

In the first half of this century very little was written about Molyneux's problem. Most publications linked the problem to the empiricism-nativism debate or to a variety of accounts of congenitally blind patients cured by surgery, as was the case in an earlier period.[1] Interest in Molyneux's problem can be seen to increase from the Second World War onwards. The interest was mainly historical, in that various biographers of and commentators on philosophers for whom the problem was of major importance (such as Locke, Berkeley, Condillac and Diderot) gave their views on the solutions put forward by these philosophers. To the extent that these authors advanced interesting viewpoints, they have been mentioned in previous chapters. The problem is also regularly discussed in textbooks or general histories of psychology, ophthalmology, neurophysiology and the like.[2]

A few researchers showed particular interest in the course of the discussion on Molyneux's problem and have written a more or less summary history of it. John Davis, for instance, in his article "The Molyneux Problem" (1960) gave a brief summary of the major eighteenth-century and of some nineteenth-century participants in the debate. He came to the conclusion that "the problem illustrates a phenomenon which has occurred more than once in the history of ideas—the shift of a problem from a speculative philosophical issue through a phase as a psychological problem to an ending as a problem for the

[1] See, for example, Bourdon 1902, Schlodtmann 1902, Villey 1914, Dennis 1934, and Murray 1944.

[2] See, for example, Klein 1970, Pastore 1971, Murray 1983, Dember & Bagwell 1985, Van Hof & Walter [1978] 1984, Crone 1992, and Zeki 1993.

physiologist."[1] In his book *Molyneux's Question* Michael Morgan discussed a few (mainly eighteenth-century) answers and dealt with a number of matters which have to do with Molyneux's question only indirectly. Morgan's view is interesting: the problem, he stated, requires a new approach in view of the fact that three hundred years of discussion have failed to lead to a definitive solution. Morgan's suggestion that blind people should be offered information viâ what is known as a sensory substitution system will be discussed below.

In addition to historical analyses it is also possible to find fresh attempts to solve Molyneux's question. First of all, use has been made of information on the visual capacity of blind people who have been operated on, just as happened in the eighteenth and nineteenth centuries.[2] What is new is that the results of animal experiments involving deprivation have been related to Molyneux's problem. The same applies to the use of sensory substitution systems.

2 Surgically Treated Cataract Patients

Over the centuries cataract has been diagnosed and treated at increasingly earlier stages, leading to a decrease in the number of reports of (older) patients operated on. The reports which are published hardly differ from the cases we have already examined. The fact that a couple of them are discussed here is merely for the sake of completeness.

By far the most influential was Von Senden's *Raum- und Gestaltauffassung bei operierten Blindgeborenen* (1932) which contains reports on sixty cataract operations. Von Senden stated that Molyneux was the first to draw attention to the importance of such patients for theories regarding our conceptions of space. Using various cases Von Senden attempted to investigate whether the tactile impressions of blind people provide them with any spatial awareness and he also wished to discover how spatial awareness develops in blind people who have been surgically treated.

Like Platner, as already stated, Von Senden believed that spatial awareness can only be obtained through sight. He came to this conclusion "because it repeatedly emerged throughout virtually all the cases, that everything spatial presented to the patient after the operation is entirely new to him, and that no bridge, however narrow, can be opened up to it from his tactual mode of

[1] Davis 1960, p. 408.

[2] Information on the faculty of sight in newborn animals and babies has not played any significant rôle in the discussions surrounding Molyneux's question in the twentieth century.

existence."[1] In fact, however, he took as proven what had to be proved: he assumed that a real awareness of space can only be visual.[2]

According to Von Senden, immediately after the operation the patients were able to distinguish objects one from the other and could localise them at a certain distance, but they could not identify any objects. When it appeared that some could, indeed, identify objects, he attributed it to the fact that they were using other senses or had prior knowledge of the situation.[3]

In general Von Senden's book was warmly received because of the richness of the material it contained, but his conclusions were not always equally appreciated. The neuropsychologist Donald Hebb regarded Von Senden's work as proof of his empirical theory of the perception of shapes. The various cases showed, he thought, that while there is an inborn *figure-ground* mechanism, experience is required in order that an object be seen as a coherent whole.[4] Other researchers were more critical, however: they pointed to the fact that Von Senden's material is frequently not amenable to unambiguous interpretation and rejected his statement that the blind have no concept of space.[5]

The psychologists Richard Gregory and Jean Wallace criticised Von Senden and Hebb on the grounds of their own study of a blind person who had been operated on. Viâ a local paper they had learnt that a fifty-two-year-old man who had become blind at the age of ten months had recovered his sight after an operation. According to the doctor in question the patient had recognised day-to-day objects, such as tables and chairs—immediately after the operation.

Gregory and Wallace tracked the patient down and began their study forty-eight days after the operation on the left eye. (The right eye was operated on three weeks after the left.) They published their findings in a monograph entitled *Recovery from Early Blindness* (1963).[6] The fact that the man was able to distinguish by sight capital letters he had learnt by touch but failed to do so with lower case letters that he had never felt they regarded as the main result of their study.[7] Gregory and Wallace believed that this indicated

[1] Von Senden [1932] 1960, p. 309.

[2] See, for example, Von Senden [1932] 1960, p. 66 and p. 106.

[3] Von Senden [1932] 1960, p. 117. Diderot [1749] 1961 and Abbott 1904 had already indicated this.

[4] Hebb 1949, pp. 19–21 and pp. 26–31, respectively. Hebb's theory was, as far as the second point is concerned, the opposite of the *Gestalt* theory. See Köhler 1947, p. 149: "The elementary nature of continuous wholes is demonstrated by observations on the first reactions of congenitally blind adults who see after an operation."

[5] See, for example, Zuckerman & Rock 1957; Warnock in Von Senden [1932] 1960, pp. 319–325; Jones 1975; Evans 1985.

[6] Reprinted in Gregory 1974a.

[7] Gregory & Wallace 1963, p. 40. Valvo 1971 appears to come to similar conclusions. See Gregory 1974a, pp. 67–68. I was unable to consult Valvo's book myself.

a transfer of perceptual information from touch to sight.[1] They were also of the opinion that the case histories of cataract patients can tell nothing of the nature of perceived space, as Von Senden had hoped, and they further doubted whether such cases could throw light on the normal development of the faculty of sight in young children, as Hebb had believed. The difficulties experienced by many of the patients mentioned by Von Senden in learning to see could, they said, be ascribed to the patients' long reliance on the sense of touch.[2]

The results obtained by Gregory and Wallace stand in stark contrast to what Ackroyd, Humphrey and Warrington reported in their article "Lasting Effects of Early Blindness: A Case Study" (1974). Ackroyd, Humphrey and Warrington studied a woman who had lost her sight at age three and underwent a successful operation on her left eye when she was twenty-seven. Six months after the operation she was able to follow the movement of prominent objects and could avoid certain obstacles, but she was totally unable to recognise objects from their shape. Eighteen months after the operation she resumed her life as a blind person.[3]

A possible explanation of the results suggested by Ackroyd, Humphrey and Warrington was that the visual cortex of the patient had not received sufficient stimulation in the critical period and that certain cells (known as feature analyzers) were thus absent or abnormal. The researchers compared the remaining seeing capacity of the patient with that of apes whose primary visual cortex has been removed. However, in view of the fact that disturbances in identification also occur with temporal lesions, it is not all that clear where the deviation should be looked for in the nervous system. The deviations do not necessarily have to occur in the visual cortex. Ackroyd, Humphrey and Warrington further suspected that the patient studied by Gregory and Wallace had had more pre-operative experience than was admitted.

The above shows that modern reports on congenitally blind people (or people who became blind at an early age) operated on successfully provide a picture which is just as ambiguous as was the case in previous centuries.

---

[1] In a later article Gregory stated that he had shown a sphere and a cube to the patient immediately after the operation and that the man had named them correctly as round and square. However, the original report does *not* mention this experiment. Gregory also suggested that Molyneux's question had distracted research from what he considered to be fruitful territory, namely the study of the "innate" knowledge of babies. See Gregory 1974b, p. 429. See also Mackie 1974 and Berman 1974b.

[2] Gregory & Wallace 1963, p. 39: "It would seem that the difficulty is not so much in learning *per se* as in *changing perceptual habits and strategies from touching to seeing*."

[3] Sacks 1993 describes a patient who "had at first been unable to recognize any shapes visually [...]. To him, a touch square in no sense corresponded to a sight square. This was his answer to the Molyneux question."

The same applies to the interpretation of the case studies. The various cases are difficult to compare because of the differences in pre-operative and post-operative situations. In addition it is often unclear from what age and to what extent the person in question was blind. The only progress made is that greater insight has been obtained into the effects of long-lasting (total or partial) light deprivation on the development of the visual system.

Despite the difficulties, the various reports are used as evidence for solutions to Molyneux's problem. A few people have regarded the results as confirmation of Locke's position.[1] But most researchers have concluded that a congenitally blind person undergoing a successful operation can indeed distinguish a sphere from a cube but cannot name the objects correctly.[2] Gregory and Wallace were the only ones to show conviction that the patient in question was also able to name the objects correctly.

A number of authors have expressed criticism. They regarded the contradictory results unsuitable for providing an answer to Molyneux's question. Morgan, for instance, wrote the following: "We cannot answer confidently because the evidence is contradictory."[3] John Heil expressed himself in similar terms: "evidence bearing on Molyneux's question remains inconclusive and unsatisfactory."[4] Crone too believed that cataract patients can provide no conclusive evidence because they have forgotten the skill of seeing thanks to their long-term blindness.[5]

To avoid the problems surrounding the clinical cases, alternatives with greater experimental control were sought. These included animal experiments involving deprivation. In 1970 the historian Klein thus wrote: "In recent decades the [Molyneux] problem has been tackled by animal physiologists by methods more amenable to *experimental* control than is possible in *clinical* studies."[6]

[1] See, for example, Klein 1970, p. 391, and Dember & Bagwell 1985, pp. 279–280.

[2] See, for example, Von Senden [1932] 1960, p. 169, Zuckerman & Rock 1957, pp. 286–287, Mackie 1976, p. 31, Van Hof-Van Duin 1981, p. 1655, and Van Hof & Walter [1978] 1984, pp. 524–525.

[3] Morgan 1977, p. 180.

[4] Heil 1987, p. 233.

[5] Crone 1992. See also Murray 1944, p. 607, Mackie 1976, p. 31, and Evans 1985, pp. 380–382.

[6] Klein 1970, p. 391.

3 Visual Deprivation in Animals

Since Spalding's experiments with newborn chicks deprived of light for a few days by means of a small cap, various species of animals have been the subject of studies into the effects of visual deprivation on the development of the visual system.[1] These animals have included fish, rats, rabbits, cats and chimpanzees, allowed to mature in variable periods of total darkness from birth. Once the period of darkness had passed, the eyes of the animals were exposed to a quantity of light, limited or not as the case may be, and the results of the deprivation were studied systematically. The results showed that the effects of light deprivation vary from species to species.[2]

Various researchers have used the results of deprivation experiments to solve Molyneux's problem. In view of the fact that such experiments performed on human beings are generally unacceptable from a moral viewpoint (despite the plea made by the eighteenth-century philosopher Mérian) and animal experiments seem to meet with less resistance, researchers have based their conclusions mainly on the experiments performed on chimpanzees, Man's close relative.[3]

In the forties the psychologist Austin Riesen allowed two chimpanzees to spend the first sixteen months of their life in darkness. After the period had passed they exhibited reflexes which showed that their eyes were sensitive to light. However, they were incapable of recognising by sight objects—such as a feeding bottle—that they had learnt to recognise by touch. After the sixteen months of darkness Riesen exposed one of the chimpanzees to a limited amount of light for five months and the other for seventeen months. Then the animals were placed in an environment with normal daylight. After a number of months the first chimpanzee managed to recognise objects, but the behaviour of the second appeared to be greatly retarded. At first its faculty of sight improved briefly but then relapsed rapidly.[4]

In another experiment a chimpanzee that had spent the first seven months of its life in a normal environment and had good eyesight was closed up in darkness until the age of twenty-four months. Subsequently it was unable to recognise its feeding bottle and other objects and was even unable to look at people or objects. Its recovery was slow and partial.[5]

---

[1] See Spalding 1873.

[2] See, for example, Beach & Jaynes 1954.

[3] See, for example, Riesen 1950, p. 17: "Obviously such an experiment cannot be risked in human beings; [...] The most logical subject for the experiment is another higher primate."

[4] Riesen 1950, pp. 16–19.

[5] Riesen 1950, p. 19.

In a third experiment Riesen had a chimpanzee spend the first three months of its life in darkness. This animal took longer to react adequately to visual stimuli than did chimpanzees that had spent the first seven months in darkness. Riesen believed that this indicated that visual discrimination is a matter of both maturity and learning.[1]

Later research showed that long-term stimulus deprivation had led to degeneration of some of the retinal ganglion cells.[2] During the development of the visual system there seems to be a particular period of sensitivity to deprivation of light. Within this sensitive or critical period changes brought on by deprivation can be undone. But when exposure to light is postponed for too long the development of normal visual mechanisms becomes extremely difficult if not impossible.

On the basis of other experiments Riesen came to the conclusion that it is not only *light* itself that is essential for normal visual development: stimulus through visual *patterns* is also an essential requirement.[3] Other researchers confirmed this. The famous studies carried out on cats by Hubel and Wiesel showed that individual neurons react to simple visual forms, such as a thin moving rod in a particular orientation against a constant background surface.[4] Neuronal detectors of this type are present at birth and their behaviour can be modified by experience. Blakemore and Cooper, for instance, demonstrated that cats raised in an environment lacking vertically oriented stimuli have a reduced number of neurons that react to vertical stimuli. The animals showed perceptual deviations: they collided with such objects as the legs of tables and chairs.[5]

As already said, the results of visual deprivation experiments were again referred to in the context of Molyneux's problem. Riesen regarded his experimental results as complementary to the clinical data assembled by Von Senden.[6] The neurophysiologist Van Hof-Van Duin described Molyneux's problem as a deprivation experiment raised in order to establish a point of view in the empiricism-nativism debate. She was of the opinion that although it was impossible to take up an unambiguous position, modern deprivation experiments had clearly shown that the visual system can be permanently affected under the influence of abnormal conditions during development.[7] The

---

[1] Riesen 1950, p. 19.

[2] Chow, Riesen & Newell 1957 and Riesen 1960, p. 35.

[3] Riesen 1950, p. 18.

[4] See, for example, Hubel & Wiesel 1959 and Hubel 1988.

[5] Blakemore & Cooper 1970. See also Hirsch & Spinelli 1970.

[6] Riesen 1947, p. 107: "These results can best be interpreted in conjunction with the data of Senden. Lacunæ in each set of findings, clinical and experimental, are in many respects filled by the other." See also Hebb 1949, p. 18 and p. 32.

[7] Van Hof-Van Duin 1981, p. 1657. See also Dember & Bagwell 1985, p. 285.

physiologists M. W. van Hof and W. G. Walter regarded Molyneux's question as essentially "the still current question of the extent to which neuronal elements and their mutual connections are formed under the influence of afferent action potentials."[1] They too believed that deprivation experiments had demonstrated that environmental stimuli are of major significance in the development of the visual system. Deviations in the visual system as a result of being brought up in abnormal circumstances should, they thought, not simply be interpreted as evidence of merely partially pre-coded interneuronal connections. For it is also possible that a network may have started off by following the normal course of development but has fallen into secondary degeneration as a result of insufficient functioning.[2]

The historian Klein believed that the position taken up by Molyneux and Locke had been confirmed by deprivation studies: "the general drift of the modern experimental evidence is in essential agreement with their prediction of the need for special visual experience before cataract patients can learn to recognise objects like spheres and cubes by sight."[3] However, Morgan was of the opinion that a conclusion of this sort was premature: "The evidence at present is insufficient to justify a definite decision on Molyneux's question, particularly because the animal experiments have not been specifically aimed at the problem of transfer between touch and vision."[4] By way of an alternative he suggested that the effects of the use of sensory substitution systems should be examined.

4 Sensory Substitution Systems

When, after three hundred years of discussion, no definitive solution has been found for a particular problem, said Morgan, the problem is either useless or it has been subjected to the wrong approach.[5] The discovery of a new technique or the invention of a new piece of equipment can sometimes breathe life into an exhausted problem. "It promises to be thus with Molyneux's question, which, after long and inconclusive debates about recovery from blindness, has been recently tackled by a totally new approach," was the way Morgan expressed

[1] Van Hof & Walter [1978] 1984, p. 522.
[2] Van Hof & Walter [1978] 1984, pp. 531–533.
[3] Klein 1970, p. 392.
[4] Morgan 1977, p. 191. According to Morgan cross-modal experiments provide no answer to Molyneux's question, nor do experiments with visual and haptic illusions. See Morgan 1977, pp. 192–197.
[5] Morgan 1977, p. 198.

his opinion.[1] The approach he was referring to was the use of what was called the *Tactile Visual Substitution System* (TVSS) developed at the end of the sixties by the Smith-Kettlewell Institute of Visual Sciences.[2]

A TVSS is a device which can be used to project images of objects onto the skin of a person, blind or not. Its major components are a television camera and a matrix of twenty by twenty electrically powered vibrotactors which are fixed to the back or stomach of the observer. The television camera transmits signals viâ an electronic circuit to the vibrotactors, causing some of them to vibrate, depending on the pattern registered by the camera. Each vibrotactor covers a small area of the image captured by the camera, just as a newspaper photograph reproduces a situation as a series of tiny dots. The pattern of tactile stimulation corresponds roughly to an enlarged visual image.

It turned out that provided that participants in the experiment were able to move the camera actively themselves, after practising for some time they were able to learn to distinguish and identify objects with the aid of the TVSS.[3] Bach-y-Rita conducted experiments involving blind people, who first were given an explanation of how the equipment worked and were then trained to distinguish between horizontal, vertical, diagonal and curved lines. Then they learnt to recognise combinations of lines (such as circles, squares and triangles) and solid objects. After an hour's practice the leader of the experiment showed the subjects a series of day-to-day objects and drew their attention to the various components of an object, to the relationships between the components and to the object as a whole. Once each object had been analyzed the leader of the experiment presented the subjects with a particular object and asked them to identify it. At first it took five to eight minutes before an object was recognised; after ten hours' practice the reaction time was no more than a few seconds.[4]

It became clear that test subjects can learn to determine not only the shape of objects but also their position, relative size, number, orientation and direction, as also the speed of movement of the camera. They form visual concepts such as perspective, shadows, shape distortion as a function of viewpoint and apparent change in size as a function of distance.[5] Test subjects react with a shock when an object suddenly looms up. A remarkable thing is that sensations of stimuli to the skin disappear in the course of time and give way to the perception of external objects: "[the] subjects spontaneously

---

[1] Morgan 1977, p. 198.

[2] See Bach-y-Rita, Collins, Saunders, White & Scadden 1969, White, Saunders, Scadden, Bach-y-Rita & Collins 1970, and Bach-y-Rita 1972.

[3] See Bach-y-Rita 1971, pp. 281–290, and Guarniero 1974.

[4] Bach-y-Rita 1971, p. 283.

[5] Bach-y-Rita, Collins, Saunders, White & Scadden 1969, p. 963.

report the external localization of stimuli, in that sensory information seems to come from in front of the camera, rather than from the vibrotactors on their back."[1] The vibrations are perceived as no more than itching as far as discomfort is concerned. Guarniero, one of the blind test subjects, wrote that "the experienced quality of the sensations was nothing like that perceived by touch. [...] I never discovered any connection between how something 'looked' and how it felt."[2] Both sighted and blind people were seen to be capable of what came to be called "tactile vision."[3]

According to Morgan, the invention of the TVSS breathed new life into the debate on Molyneux's question. In fact he believed that it should be possible to use a TVSS to provide visual information to blind people and that in this manner the power of sight could be restored to the blind. Morgan meant this literally: "We can call it visual because the nature of the information being extracted, and the way in which it is presented, is similar to that extracted and presented by the eye [...] In general there is little doubt that the TVSS allows the blind to see."[4]

Study of the use of sensory substitution systems had, stated Morgan, demonstrated that "perception is not necessarily to be equated with the input from particular sense organs, still less with inputs along particular nerve fibres, as the obsolete law of specific energies proclaimed. Perception is the recognition of certain properties in the input."[5] Locke's answer to Molyneux's question was, believed Morgan, incorrect to the extent that it implied that only by association was it possible to recognise that two messages received along different paths actually referred to the same object. The emphasis which Locke and other empiricists placed on experience was, said Morgan, closer to the truth.[6]

Like Morgan, David Warren and Edward Strelow connected sensory substitution systems to Molyneux's problem. In their article "Learning Spatial Dimensions with a Visual Sensory Aid: Molyneux Revisited" (1984) they suggested that Molyneux's question in its pure form probably cannot be answered, but that the use of sensory substitution systems raises questions which

---

[1] Bach-y-Rita, Collins, Saunders, White & Scadden 1969, p. 964.

[2] Guarniero 1974, pp. 101–102.

[3] Gregory noted similarities between the process of learning to "see" tactually and the stages which the blind people he had operated on had passed through. See Gregory 1974a, p. 67.

[4] Morgan 1977, pp. 201–203. I would regard it as more correct to agree with Bach-y-Rita, Collins, Saunders, White & Scadden 1969, and say that the TVSS allows the blind to "see."

[5] Morgan 1977, p. 207.

[6] Morgan 1977, p. 207.

resemble Molyneux's and can, indeed, be resolved. They regarded the use of such systems as an appropriate approach to the use of a new sense.[1]

Warren and Strelow set up a number of experiments to study the way in which test subjects learn to use what is known as a *Binaural Sensory Aid* (BSA).[2] A BSA is an electronic apparatus worn on the head like a cap. It consists of two components: an ultrasonic generator that transmits ultrasound and two receivers which collect the echoes of the ultrasound and translate them into audible signals. (A difference between a TVSS and a BSA is that a TVSS turns visual information into tactile information while a BSA relates exclusively to hearing and provides sonic information in a form perceptible to hearing.) The equipment is designed in such a way that the echoes provide information on the distance, the direction and the surface structure of objects. The frequency of the sound reflected specifies the distance; the interaural amplitude difference indicates the direction; and the clarity of the signal is indicative of the structure.

Experiments which Warren and Strelow carried out using blindfolded people showed that some time was required for the test subjects to learn to determine distance and direction: "Much as Locke predicted for the use of visual information by the person newly recovered from blindness, we found in these experiments that there was no immediate understanding of the information provided by the sensors."[3] However, the time required was not all that long, which Warren and Strelow believed could indicate that a blind person gaining the power of sight would be able to learn a considerable amount in a short time if, as in the case of the blindfolded test subjects, there was sufficient opportunity for training and there was enough feedback.[4] Warren and Strelow were of the opinion that there was no better approach to Molyneux's problem.

---

[1] Warren & Strelow 1984, pp. 331–332. Warren and Strelow were aware that a "sensory aid is not, strictly speaking, a new modality."

[2] The BSA was designed by L. Kay. See Kay 1974.

[3] Warren & Strelow 1984, p. 348. Heil regarded information on the use of sensory substitution systems as disproving Locke's associationist position. See Heil 1987, p. 239.

[4] Warren and Strelow also pointed out that certain results would perhaps not be obtained if blind rather than blindfolded people were used as test subjects. Research carried out by Aitken and Bower (1982) also showed that blind babies quickly observe objects rather than echoes.

5 Electrical Stimulation of the Visual Cortex

In an article entitled "Molyneux's Question" (1985) Gareth Evans proposed a variant on Molyneux's problem which, at first sight, seemed to provide a better approach to the problem than that outlined above.[1] He suggested using direct electrical stimulation in the visual cortex of a congenitally blind person to produce a pattern of phosphenes in the shape of a square.[2] Evans expected that the blind person would be able to apply to the new situation the concept of "square" that he had previously learnt to apply through the sense of touch.

As far as I have been able to discover, experiments of this nature were never performed on persons born blind. But there is no reason to expect that they would bring us any closer to a solution of Molyneux's problem than would tests carried out on cataract patients, since in both cases the visual system has undergone abnormal development.

6 Conclusions

As was the case in the nineteenth century, Molyneux's problem played not such a prominent rôle in the twentieth century as it did in the eighteenth century. And yet it would appear that it still keeps minds occupied. Many modern publications on the subject are of an historical nature, usually interpretations of the positions taken up by famous philosophers in earlier times. A couple of authors have written summary histories of the problem. In addition philosophers, psychologists and other scientists have themselves made attempts to solve the problem, using tools both old and new. The old methods included appeals to results obtained in tests carried out on congenitally blind people whose sight has been restored surgically. The approach provided no new insights, particularly because of the problems of interpretation, which refuse to become less with the passage of time.

More interesting are the results of experiments in which animals were deprived of visual stimuli. Among the data such experiments have provided is the knowledge that the nature of the stimulus during what has come to be called the critical period is of importance in the development of the visual

[1] In this interesting article, Evans aimed to show "how Molyneux's Question is linked to fundamental issues in the philosophy of mind and perception" (p. 399). See also Campbell 1989.

[2] Evans 1985, pp. 392 ff. See also Brindley & Lewin 1968 and Dobelle 1977.

system. Experimental animals raised in a dark environment or one providing a limited number of visual stimuli acquired abnormal powers of sight. Deprivation experiments are, in fact, unsuitable for solving the original problem posed by Molyneux, since he seems to have assumed that congenitally blind people recovering the faculty of sight would immediately be endowed with normal, properly functioning vision. To the extent that researchers have insisted on using the results of deprivation experiments as direct evidence for a solution to Molyneux's problem, they have ended up defending the position of Molyneux or Locke.

Deprivation experiments have thrown new light on the visual behaviour of patients operated on for cataract. Such patients do, in fact, have some perception of light but none of shape. This leads to their visual system developing abnormally. Depending on the age at which the cataract develops and is treated and depending on the extent of the blindness, cataract patients can learn to use their visual system to a varying extent after a successful cataract operation. This explains the divergent results of the various cases. For cured cataract patients the same applies: they are not appropriately qualified to help in solving Molyneux's problem since their powers of sight, in contrast to that of Molyneux's congenitally blind person, is abnormal. The same would seem to apply to experiments involving direct electrical stimulation of the visual cortex of congenitally blind people.

The most recent approach to Molyneux's problem involves the use of sensory substitution systems which were first and foremost developed to increase blind people's mobility and to teach them to read. Learning to use sensory substitution systems such as a TVSS or a BSA is regarded as a good approach to Molyneux's problem since equipment of this sort creates an opportunity for seeing how test subjects react to a new type of sensory information. Experiments conducted using sensory substitution systems have led researchers to conclude that time is needed for the subject to learn how to use "new" sensory systems, and they have proposed this conclusion as evidence supporting Locke's position. And although this is, indeed, no more than an *approach* to Molyneux's problem, we should not expect a better one to be found.

CHAPTER SEVEN

MOLYNEUX'S PROBLEM IN RETROSPECT

1 INTERPRETATIONS OF MOLYNEUX'S PROBLEM

IN the foregoing chapters we have made an extensive examination of the history of the discussion surrounding Molyneux's problem and thus the aim of this book is achieved. To close the examination, we will summarise the various interpretations and approaches proposed over the course of three centuries with reference to this problem.

When he set out his problem Molyneux formulated a number of conditions and assumptions which were not completely clear or acceptable to everybody. This led inevitably to varying interpretations of the problem.

One of Molyneux's assumptions was that an *adult congenitally blind person* had learnt by touch to distinguish a sphere from a cube and had also learnt to name them correctly. Most people regarded this as a plausible assumption, since practically everyone believed that the congenitally blind can acquire *knowledge of geometric properties of objects.* A person born blind could, therefore, learn that a cube has angles whereas a sphere does not. He would also be capable of learning that it is impossible to see all sides of an object at the same time and that the visual appearance of anything depends on the angle of vision. According to most philosophers, such as Leibniz and Reid, a person born blind would also be capable of learning that the geometric properties that he has learnt viâ touch correspond in one way or another to the geometric properties which the sighted learn to know viâ sight.[1] The fact that a blind man like Saunderson could practise optics and geometry reinforced them in their belief. Only Platner and his disciples held the opinion that the congenitally blind can form no concepts whatsoever of space or geometric properties.

Diderot suggested that it was important whether the blind person in question was *intelligent* or not. An uneducated person would, he thought, be incapable of arriving at the knowledge detailed above. A number of eye specialists

[1] As we have seen, Berkeley disagreed with this; the early Condillac suggested that it was possible that an object resembling a sphere when seen could turn out to be a cube when felt.

and philosophers agreed with this suggestion. Diderot further remarked that the solutions proposed for Molyneux's problem could not simply be regarded as universally valid. For if a person cured of blindness were to be shown complex objects, such as a glove or a bathrobe, he would have greater difficulty identifying them. This is an acceptable assumption.

One of the conditions set by Molyneux was that the *sphere* and the *cube* offered to the blind person first to be touched and then, later, to be viewed should be made of the *same material* and should be of the *same size*. Clearly Molyneux wished to avoid the objects being recognised on the basis of their structure and size rather than their shape. Condillac and Diderot regarded such conditions as superfluous.

A further requirement was that both objects should be placed *on a table*. Although nobody suggested as much, it is not difficult to imagine this condition rendering recognition of the objects more difficult: the blind person now sighted would have to distinguish not only the cube from the sphere but also the objects themselves from the table and the table from the background. If the cube and sphere were to be placed against a plain background this problem would be obviated.[1] Most researchers assumed implicitly that the person in question would see the cube and sphere as separate objects.

One of the major assumptions made by Molyneux was that the blind person in question would have his *sight restored*. When the question was first formulated participants accepted this hypothesis for the sake of the discussion without really believing that restoration of sight was possible. All this was changed when Cheselden restored the faculty of sight to a patient with congenital cataract. From then on, many philosophers believed that the experiences of surgically treated patients with congenital cataracts were, at least in principle, suitable for solving Molyneux's problem. Only Mérian made an objection to using cataract patients for this purpose: they are never completely blind. He also pointed out that people coming into the world completely blind can never acquire the faculty of sight.

There was some lack of clarity regarding the question of whether Molyneux would *tell* the blind person that he was to be shown a sphere and a cube or not. Leibniz and Jurin, for example, assumed that the blind man would be given this information and that he would thus only be called on to make the required distinction. They believed that if this information was not imparted, the subject would be incapable of distinguishing the sphere from the cube. Others, including Condillac, thought it desirable not to ask any leading questions and therefore not to tell the man what objects he was to be shown. And indeed

[1] Diderot proposed a variant on Molyneux's problem in which a black cube and a red sphere would be placed against a white background. See Diderot [1749] 1961, p. 137.

we should expect the identification of the objects to be easier if the subject had prior information about them.

A number of philosophers, the first being Boullier, believed that it would be better to use a *circle* and a *square* rather than a sphere and a cube because—so thought these philosophers—we can only perceive (flat) surfaces. Even quite recently Evans proposed using this simplified version of the problem.[1]

One of the phrases in Molyneux's statement of his problem which was subject to a variety of interpretations was whether the formerly blind person could distinguish and name the objects *by sight* before touching them. Some philosophers were of the opinion that the subject would have to give an *immediate* reaction, as soon as his faculty of sight was restored. Others believed that he ought to take his *time*, that he should be allowed to walk around the table on which the objects had been placed and should also be allowed to use his reason and his memory, as Jurin wrote. A number of researchers saw a difference in the man being allowed only to look or also to use mathematical reasoning. Although this difference is difficult to specify, it is possible to imagine that it could matter whether the subject had to make an immediate judgment or not. Probably what Molyneux was thinking of was whether the man would be able to distinguish and name the objects without recourse to the sense of touch. Doubtless he would have been given some time to attempt this. Many philosophers actually expected that he would be too amazed initially to say anything at all.

As already stated, Molyneux's problem does, in fact, consist of two questions: can a person born blind who has learnt to distinguish a sphere from a cube by touch and to name both objects *distinguish* them by sight alone on restoration of that faculty and, further, can he *name* them correctly? Molyneux himself—and many others—failed to treat these two questions separately: they gave only a single answer. This is not surprising when, by answering "no" they were indicating that they did not believe the man would be capable of distinguishing the sphere from the cube, for in that case he would certainly not be capable of naming the objects. But if a negative answer referred only to the subject's inability to name the objects, it is then unclear whether they also refused to accept that he would be able to distinguish the sphere from the cube or whether they thought that he would succeed. Failure to name the objects in no way implies failure to be able to distinguish them. (A positive answer usually produced no confusion because it generally referred to the ability to *name* the objects and thus also accepted that the formerly blind person would be able to distinguish the sphere from the cube.) It is thus important that both questions be answered separately.

[1] Evans 1985, p. 365.

2 Ways of Dealing with Molyneux's Problem

Partly as a result of developments within various branches of science, Molyneux's problem has been dealt with in various ways in the course of the centuries and a variety of solutions have been proposed grounded in diverse arguments. Originally the approach was *speculative*. Philosophers made attempts to get at the link between visual and tactile sensations of the shape of objects and they wondered whether the link could be directly observed or not. They also attempted to discover whether the concepts of form obtained in one way or another viâ the sense of touch are the same as or have something in common with the concepts of form obtained through the faculty of sight—and if so, whether this could be immediately noted.

There was agreement that visual and tactile sensations differ from one another qualitatively. However, opinion was divided over the link between visual and tactile sensations of the form of objects. Some philosophers were of the opinion that there was a necessary relationship between the two. Some of them believed that the relationship could be discerned immediately, others thought that it came through experience and still others put it down to (mathematical) reasoning. Other philosophers believed that there was no essential connection between the two and that experience created no more than an arbitrary link.

With regard to the relationship between tactile and visual concepts a variety of opinions were defended. Some philosophers were of the opinion that the tactile and visual concepts of a sphere and a cube are essentially the same, or at least show similarities, and they believed that this identity or similarity could be observed either immediately or with the aid of reason. Others held that tactile and visual concepts differ radically from one another and that experience is required to create a link between the two types of concept.

The question of how concepts are formed was usually left unanswered. Only very few attempted to clarify the links that could exist between sensations and concepts. Hutcheson and Boullier believed, for instance, that concepts and sensations accompany one another; and Mérian believed that concepts can be abstracted from sensations. Leibniz pointed out that a perceiver may have a sensation of an object without having a clear concept of that object and that the possession of an exact concept of two objects does not imply that these objects can be immediately distinguished on sight.

After the publication of Cheselden's account, it was believed that the various positions could be assessed *empirically* as to their correctness. (Mérian was one of the few who questioned this.) The fact that the cataract patient operated on by Cheselden was at first unable to distinguish objects of any shape or size from one another demonstrated, some believed, that Molyneux's question had to be answered in the negative. Many of these people were to be found in Great Britain, especially among followers of Berkeley's theory of vision. However, those who had formed an opinion, often based on speculative grounds, that a congenitally blind person whose sight is restored would be capable of distinguishing a sphere from a cube and would be able to name the objects correctly, stated that Cheselden's study had not been performed in an adequate manner. The philosophers in question, mainly French *philosophes*, provided alternative explanations for the phenomena observed and drew up criteria governing a reliable preparation and style of questioning of cataract patients who were to be operated upon.

Later case studies of such patients, however, turned out to involve just as many problems, thereby providing no answer to Molyneux's question. One of the reasons for this was that pre-operative and post-operative conditions differed from case to case. Prior to the operation the patients suffered from different degrees of blindness and had widely differing intellectual backgrounds. Furthermore the doctors failed to conduct their experiments at the same time following the operation and the experiments were not identical. As already noted, Mérian expressed fundamental criticism of the use of cataract patients; and Diderot stated expressly that he attached more importance to the judgment of a philosopher with common sense than to the statements of a congenitally blind person who had just undergone surgery. Despite the difficulties listed, case studies of cataract patients treated surgically continued to be used as proofs of solutions of Molyneux's problem.

The problems mentioned above are among the reasons why, in the nineteenth century, other types of empirical evidence were introduced into the discussion on Molyneux's problem. This material consisted largely of data about the visual capacities of newborn animals and babies. There was agreement on the fact that certain animals are capable of distinguishing objects one from the other and of estimating distance immediately after birth. There was, however, no agreement on whether this also applied to babies. The solutions proposed to Molyneux's problem on the basis of these findings related to the point of view adopted with regard to the extent to which the faculty of sight in newborn animals and human babies could be compared to that of congenitally blind adults cured of their blindness.

Study of the visual behaviour of young animals and neonates was interesting not only in reference to Molyneux's problem but, particularly, for theories of sight in general. For instance, the fact that animals can distinguish objects and see depth independently of the sense of touch proved—just like Wheatstone's explanation of stereoscopic vision—that Berkeley's theory of vision was untenable.

In the nineteenth century Molyneux's question was also answered on the basis of other grounds. A number of researchers pointed out that sighted people have difficulty recognising complicated or unusual shapes when using their sense of touch. By way of analogy, they believed that a congenitally blind person recovering his sight would have similar problems in recognising a sphere or a cube. As Diderot had already suggested, this argument may apply to complicated objects but the question is whether it also applies to objects as geometrically simple as spheres and cubes.

In the twentieth century Molyneux's problem was approached in two more ways. In the first place knowledge about the effects of visual deprivation in animals was applied to the problem. Experiments showed that the visual system changes under the influence of abnormal conditions during development. If, during a certain critical period, an animal is not stimulated by visual patterns, it will never learn to distinguish objects one from the other—or will do so only with the greatest difficulty.

Finally, the most recent approach to Molyneux's problem has been the use of sensory substitution systems. Such systems provide information which is coded in a special way and is normally obtained viâ another route. It is not the case that the sensations normally acquired are of a similar nature to those acquired by a sensory substitution system. And it is therefore incorrect to join Morgan in stating that blind people really can see with the aid of a Tactile Visual Substitution System. Warren and Strelow rightly noted that the use of sensory substitution systems merely *approaches* the use of a new sense and that research into the operation of equipment of this type cannot, therefore, provide a definitive solution to Molyneux's problem. To the extent that the conditions under which users of sensory substitution systems operate are analogous to those in which Molyneux's hypothetical subject finds himself, the results of the experiments are interesting. Experiments have demonstrated that test subjects need time to learn to distinguish objects and to name them. It has also been noted that feedback improves this process.

In view of the fact that we can hardly expect that a more direct approach to the problem will be found than the alternatives we have mentioned, the history of Molyneux's problem as outlined can be regarded as a closed book.

3 Conclusions

We have not answered Molyneux's question—and, indeed, we think that it cannot be answered because congenitally blind people cannot be made to see once their critical period is passed. The question was, however, worth putting: it led to discussions and experiments which have provided us with a veritable treasure house of information. In the course of history the question has, in fact, been replaced by numerous other questions which, while related to the original problem, have gone on to lead a life of their own. Some of these are philosophical in nature, such as problems relating to the conceptual clarification of perceptual terms, to the justification of claims to knowledge, to the status of philosophical thought experiments,[1] and so on. Others are in the field of psychology; for example, psychological questions concerning pattern recognition, learning processes, formation of concepts, methodological issues, and so on. Still other questions, such as those concerned with the study of the development of the visual system, critical periods for development, and the nature and function of the association cortex belong to the domain of neurophysiology.

And thus the discussions surrounding Molyneux's problem provide us with the same image as we get from a river. The problem had its source in 1688 in Molyneux's mind. It bubbled along a little until Cheselden performed his cataract operation in 1728 and initiated a flood. The question then quit the high country of philosophy and flowed into the wide plains of experimental psychology and neurophysiology. The mighty stream into which it then flowed has since spread out to form such a wide delta that it is no longer possible to measure its extent.

[1] Wilkes 1988 devotes some attention to the often problematical status of philosophical thought experiments.

BIBLIOGRAPHY

ABBOTT, E. A. 1884. *Flatland: A Romance of Many Dimensions.* London: Seeley and Co.

ABBOTT, T. K. 1864. *Sight and Touch: An Attempt to Disprove the Received (or Berkeleian) Theory of Vision.* London: Longman, Green, Longman, Roberts, and Green.

ABBOTT, T. K. 1904. "Fresh Light on Molyneux' Problem: Dr. Ramsay's Case." *Mind* **13**, 543–554.

ACKROYD, C., HUMPHREY, N. K. & WARRINGTON, E. K. 1974. "Lasting Effects of Early Blindness: A Case Study." *Quarterly Journal of Experimental Psychology* **26**, 114–124.

AITKEN, S. & BOWER, T. G. R. 1982. "The Use of the Sonicguide in Infancy." *Visual Impairment and Blindness* **76**, 91–100.

ALEMBERT, J. LE ROND D'. 1751. "Aveugle." In: Diderot & D'Alembert (eds.) 1751–1780, vol. 1 (1751), pp. 870–873.

AMMAN, J. C. 1692. *Surdus loquens seu methodus quâ qui surdus natus est loqui discere possit.* Amsterdam.

ANTI-BERKELEY. 1752. "Errors in the *New Theory of Vision* Supposed to have been Wrote by Dr. *Berkeley*, the Present Bishop *of Cloyne*: First Publish'd in 1709, and Afterwards with the *Minute Philosopher* in 1732." *Gentleman's Magazine* **22**, 11–13.

APELT, E. F. 1857. *Metaphysik.* Leipzig: Wilhelm Engelmann.

ARISTOTLE. 1968. *De Anima.* In: *Aristotle's De Anima: Books II and III.* Translated with Introduction and Notes by D. W. Hamlyn. Oxford: Clarendon Press, 1968.

ARNOBIUS OF SICCA. 1949. *The Case against the Pagans.* Newly Translated and Annotated by George E. McCracken. Westminster, Maryland: Newman.

BACH-Y-RITA, P. 1971. "A Tactile Vision Substitution System based on Sensory Plasticity." In: T. D. Sterling, E. A. Bering, S. V. Pollack & H. G. Vaughan (eds.), *Visual Prosthesis: The Interdisciplinary Dialogue.* New York: Academic Press; pp. 281–290.

BACH-Y-RITA, P. 1972. *Brain Mechanisms in Sensory Substitution.* New York: Academic Press.

BACH-Y-RITA, P., COLLINS, C. C., SAUNDERS, F. A., WHITE, B. W. & SCADDEN, L. 1969. "Vision Substitution by Tactile Image Projection." *Nature* **221**, 963–964.

BAER, K. E. VON. 1824. *Vorlesungen über Anthropologie, für den Selbstunterricht bearbeitet.* Königsberg: Gebrüder Bornträger.

BAILEY, S. 1842. *A Review of Berkeley's Theory of Vision, Designed to Show the Unsoundness of that Celebrated Speculation.* London: James Ridgway.

BAILEY, S. 1843. *A Letter to a Philosopher, in Reply to some Recent Attempts to Vindicate* Berkeley's Theory of Vision *and in Further Elucidation of its Unsoundness.* London: James Ridgway.

BAIN, A. 1855. *The Senses and the Intellect.* London: John W. Parker and Son.

BARROW, I. 1674. *Lectiones opticæ.* London.

BEACH, F. A. & JAYNES, J. 1954. "Effects of Early Experiences upon the Behavior of Animals." *Psychological Bulletin* **51**, 239–263.

BEER, G. J. 1813. *Das Auge, oder Versuch das edelste Geschenk der Schöpfung vor dem höchstverderblichen Einfluß unseres Zeitalters zu sichern.* Wien: In der Camesinaschen Buchhandlung.

BELL, C. 1833. *The Hand: Its Mechanisms.* London: W. Pickering.

BERKELEY, G. [1707–1708] 1975. "Philosophical Commentaries: Berkeley's Notebooks of 1707–1708." First published in: Berkeley 1871. Reprinted in: Berkeley 1975, pp. 251–336.

BERKELEY, G. [1709] 1975. *An Essay towards a New Theory of Vision.* Dublin: Printed by Aaron Rhames for Jeremy Pepyat, 1709. Modernized reprint in: Berkeley 1975, pp. 1–55.

BERKELEY, G. [1710] 1975. "An Appendix." In: Berkeley, *An Essay towards a New Theory of Vision.* Second Edition. Dublin: Printed by Aaron Rhames for Jeremy Pepyat, 1710. Modernized reprint in: Berkeley 1975, pp. 56–59.

BERKELEY, G. [1733] 1975. *The Theory of Vision Vindicated and Explained.* London: Printed for J. Tonson, 1733. Modernized reprint in: Berkeley 1975, pp. 231–250.

BERKELEY, G. 1734. *The Analyst: Or a Discourse addressed to an Infidel Mathematician wherein it is Examined whether the Object, Principles, and Inferences of the modern Analysis are more Distinctly Conceived or more Evidently Deduced than Religious Mysteries and Points of Faith.* By the Author of the *Minute Philosopher*. London: Tonson.

BERKELEY, G. 1735. *A Defence of Free-Thinking in Mathematics: In Answer to a Pamphlet of Philalethes Cantabrigiensis, intituled* Geometry no Friend to Infidelity: Or, a Defence of Sir Isaac Newton and the British Mathematicians. By the Author of the *Minute Philosopher*. London.

BERKELEY, G. 1871. *The Works of George Berkeley, D. D., formerly Bishop of Cloyne: Including many of his Writings Hitherto Unpublished.* With Prefaces, Annotations, his Life and Letters, and an Account of his Philosophy, by Alexander Campbell Fraser. Oxford: Clarendon Press.

BERKELEY, G. 1975. *Philosophical Works, Including the Works on Vision.* Introduction and Notes by M. R. Ayers. London and Melbourne: Everyman's Library.

BERMAN, D. 1974a. "Francis Hutcheson on Berkeley and the Molyneux Problem." *Proceedings of the Royal Irish Academy* **74**, 259–265.

BERMAN, D. 1974b. "Philosophical Prediction." *Times Literary Supplement*, 21 June, 670.

BEW, C. 1800. "Observations sur la cécité et sur l'usage des autres sens pour suppléer à la perte de la vue. Traduites de l'anglais." In: A. G. Schwenger (ed.), *Mémoires sur les aveugles, sur la vue et la vision, suivis de la description d'un télégraphe très simple.* Paris: Chez l'auteur, Amand Koenig, F. Didot, J. J. Fuchs; Amsterdam: Chez le successeur de C. N. Guérin; pp. 23–58.

BIDYADHAR, N. K. 1939. "Susruta and his Ophthalmic Operations." *Archives of Ophthalmology* **22**, 550–574.

BIDYADHAR, N. K. 1940. "Evolution of Cataract Surgery from the Pre-Historic Age up to the Present Time." *Antiseptic* **37**, 339–346.

BIDYADHAR, N. K. 1941. "Principles of Susrutian Ocular Therapy Interpreted in the Light of Modern Ophthalmic Science." *Archives of Ophthalmology* **25**, 582–628.

BLAKEMORE, C. & COOPER, G. F. 1970. "Development of the Brain depends on the Visual Environment." *Nature* **228**, 477–478.

BOLTON BRANDT, M. 1994. "The Real Molyneux Question and the Basis of Locke's Answer." In: G. A. J. Rogers (ed.), *Locke's Philosophy.* New York: Oxford University Press.

BONGIE, L. L. 1977. *Diderot's* Femme savante. Oxford: The Voltaire Foundation at the Taylor Institution. (*Studies on Voltaire and the Eighteenth Century* **166**.)

BONGIE, L. L. 1978. "A New Condillac Letter and the Genesis of the *Traité des sensations*." *Journal of the History of Philosophy* **16**, 83–94.

BONNET, C. 1760. *Essai analytique sur les facultés de l'âme.* Copenhague: Chez les frères Cl. et Ant. Philibert.

BORN, F. G. 1791. *Versuch über die ursprünglichen Grundlagen des menschlichen Denkens und die davon abhängigen Schranken unserer Erkenntniss.* Leipzig: Johann Ambrosius Barth.

BOULLIER, D. R. [1728] 1737. *Essai philosophique sur l'âme des bêtes: où l'on trouve diverses réflexions sur la nature de la liberté, sur celle de nos sensations, sur l'union de l'âme et du corps, sur l'immortalité de l'âme.* Amsterdam: François Changuion, 1728. Seconde édition revue et augmentée, à la quelle on a joint un traité des vrais principes qui servent de fondement à la certitude morale. Amsterdam: François Changuion, 1737.

BOURDON, B. 1902. *La perception visuelle de l'espace.* Paris: Schleicher Frères.

BRANDT, R. 1975. "Historical Observations on the Genesis of the Three-Dimensional Optical Picture (Gassendi, Locke, Berkeley)." *Ratio* **17**, 176–190.

BREATHNACH, C. S. 1984. "Biographical Sketches, No. 39: Molyneux." *Irish Medical Journal* **77**, 118.

BREIDERT, W. 1989. *George Berkeley: 1685–1753.* Basel: Birkhäuser.

BRINDLEY, G. S. & LEWIN, W. S. 1968. "The Sensations Produced by Electrical Stimulation of the Visual Cortex." *Journal of Physiology* **196**, 479–493.

BRISSEAU, M. 1709. *Traité de la cataracte et du glaucoma.* Paris: D'Houry.

BROWN, T. [1820] 1860. *Lectures on the Philosophy of the Human Mind.* Twentieth Edition. London: William Tegg, 1860.

BUFFON, G. L. LECLERC, COMTE DE. 1749–1789. *Histoire naturelle générale et particulière, avec la description du Cabinet du Roi.* Paris: Imprimerie Royale.

CAMPBELL, J. 1989. "Book Review of Gareth Evans: *Collected Papers.*" *The Journal of Philosophy* **86**, 156–163.

CAMPER, P. 1746. *Dissertatio physiologica de quibusdam oculi partibus.* Lugduni Batavorum: Typis Eliæ Luzac jun.

CAMPER, P. [1746] 1962. *Dissertatio optica de visu.* Lugduni Batavorum: Typis Eliæ Luzac jun, 1746. Reprinted in: Camper, *Optical Dissertation on Vision.* Facsimile of the Original Text with Translation and Introduction by G. ten Doesschate. Nieuwkoop: B. de Graaf, 1962.

CAMPER, P. [1766] 1913. *De oculorum fabrica et morbis.* Amsterdam: F. van Rossen, 1913. (*Opuscula selecta Neerlandicorum de arte medica* **2**.)

CASSIRER, E. 1907. *Das Erkenntnisproblem in der Philosophie und Wissenschaft der neueren Zeit.* Berlin: Bruno Cassirer.

CASSIRER, E. 1932. *Die Philosophie der Aufklärung.* Tübingen: J. C. B. Mohr.

CELSUS, A. C. 1935–1938. *De medicina.* Edited by W. G. Spencer. Cambridge, Mass.: Harvard University Press; London: W. Heinemann.

CHAPPELL, V. 1994. "Locke's Theory of Ideas." In: V. Chappell (ed.), *The Cambridge Companion to Locke.* Cambridge: Cambridge University Press; pp. 26–55.

CHESELDEN, W. 1713. *The Anatomy of the Human Body.* London: N. Cliff, D. Jackson and W. Innys.

CHESELDEN, W. 1728. "An Account of some Observations Made by a Young Gentleman, who was Born Blind, or Lost his Sight so Early, that he had No Remembrance of Ever Having Seen, and was Couch'd between 13 and 14 Years of Age." *Philosophical Transactions* [of the Royal Society of London] **35** (No. 402), 447–450.

CHESELDEN, W. 1733. *Osteographia, or the Anatomy of the Bones.* London.

CHILLINGWORTH, H. R. 1946. "William Molyneux: A Herald of Democracy." *Hermathena: A Series of Papers on Literature, Science and Philosophy* **67**, 60–67.

CHOW, K. L., RIESEN, A. H. & NEWELL, F. W. 1957. "Degeneration of Retinal Ganglion Cells in Infant Chimpanzees Reared in Darkness." *Journal of Comparative Neurology* **107**, 27–42.

CONDILLAC, E. BONNOT DE. [1746] 1947. *Essai sur l'origine des connoissances humaines: ouvrage où l'on réduit à un seul principe tout ce qui concerne l'entendement humain.* Amsterdam: Pierre Mortier, 1746. Reprinted in: Condillac 1947.

CONDILLAC, E. BONNOT DE. [1754] 1947. *Traité des sensations.* Paris: De Bure, 1754. Reprinted in: Condillac 1947.

CONDILLAC, E. BONNOT DE. 1947. *Œuvres philosophiques de Condillac.* Texte établi et présenté par G. le Roy. Vol. 1. Paris: Presses Universitaires de France.

CONDORCET, M. J. A. N. CARITAT, MARQUIS DE. [1773] 1847. "Éloge de Cheselden." In: Condorcet, *Éloges des académiciens de l'Académie Royale des Sciences, morts depuis l'an 1666 jusqu'en 1699.* Paris 1773. Reprinted in: *Œuvres de Condorcet.* Publiées par A. Condorcet O'Connor et M. F. Arago. Vol. 2. Paris: Firmin Didot Frères, 1847; pp. 120–122.

CONNOR, B. 1697. *Evangelium medici sive medicina mystica.* London.

COPE, Z. 1953. *William Cheselden, 1688–1752.* Edinburgh and London: E. and S. Livingstone.

CRONE, R. A. 1989. "De geschiedenis van het dieptezien." *Ogenblik* **14**, 108–114.

CRONE, R. A. 1992. *Licht, kleur, ruimte: de leer van het zien in historisch perspectief.* Houten: Bohn Stafleu Van Loghum.

DAHL, D. S. 1968. "William Cheselden (1688–1752)." *Investigative Urology* **5**, 627–629.

DANIELS, N. 1970. *Thomas Reid's Discovery of a Non-Euclidean Geometry: A Case Study in the Relation between Theory and Practice.* Harvard, Mass.: Department of Philosophy, Harvard University.

DAVIEL, J. 1748. "Lettre de M. Daviel, conseiller, chirurgien ordinaire du Roi en survivance par quartier, à M. de Joyeuse, docteur en médecine de l'Université de Montpellier, aggregé au Collège des Médecins de Marseille, et médecin des Hôpitaux des Galéres." *Mercure de France*, Septembre, 198–221.

DAVIEL, J. 1753. "Sur une nouvelle méthode de guérir la cataracte par l'extraction du cristalin." *Mémoires de l'Académie Royale de Chirurgie* **2**, 337–354.

DAVIEL, J. 1762. "Réponse de M. Daviel, chirurgien ordinaire, et oculiste du Roi, à la lettre de M. le baron de Haller, du 11 Novembre 1761, insérée dans le *Mercure de France* du mois de Février 1762, page 145." *Journal de médecine, chirurgie, pharmacie, etc.* **16**, 245–250.

DAVIS, J. W. 1960. "The Molyneux Problem." *Journal of the History of Ideas* **21**, 392–408.

DAZA DE VALDES, B. 1623. *Uso de los antojos.* Sevilla: D. Perez.

DELACAMPAGNE, C. 1987. "Diderot et le problème de Molyneux." In: F. Lafarga (ed.), *Diderot.* Barcelona: Edicións de la Universidad de Barcelona; pp. 87–94.

DEMBER, W. N. & BAGWELL, M. 1985. "A History of Perception." In: G. A. Kimble & K. Schlesinger (eds.), *Topics in the History of Psychology.* Vol. 1. Hillsdale, N. J., and London: Lawrence Erlbaum Associates; pp. 261–304.

DENNIS, W. 1934. "Congenital Cataract and Unlearned Behavior." *The Pedagogical Seminary and Journal of Genetic Psychology: Child Behavior, Animal Behavior, and Comparative Psychology* **44**, 340–351.

DESCARTES, R. 1637. *Discours de la méthode pour bien conduire sa raison, et chercher la verité dans les sciences. Plus la Dioptrique, les Météores et la Géometrie, qui sont des essais de cette méthode.* Leyde, Imprimerie de Ian Maire.

DEWEY, J. [1887] 1891. *Psychology.* Third revised Edition. New York: Harper and Brothers, 1891.

DIDEROT, D. [1749] 1961. *Lettre sur les aveugles, à l'usage de ceux qui voient.* Londres 1749. Reprinted in: Diderot, *Œuvres philosophiques.* Textes établis, avec introductions, bibliographies et notes par P. Vernière. Paris: Garnier Frères, 1961.

DIDEROT, D. [1751] 1965. *Lettre sur les sourds et muets, à l'usage de ceux qui entendent et qui parlent.* Édition commentée et presentée par P. H. Meyer. Genève: Librairie E. Droz, 1965.

DIDEROT, D. & ALEMBERT, J. LE ROND D' (EDS.) 1751–1780. *Encyclopédie, ou dictionnaire raisonné des sciences, des arts et des métiers, par une société de gens de lettres, mis en ordre par Diderot et D'Alembert.* Paris: Briasson.

DILTHEY, W. [1901] 1927. "Friedrich der Grosse und die Deutsche Aufklärung." Reprinted in: *Wilhelm Dilthey's Gesammelte Schriften.* Vol. 3. Leipzig und Berlin: B. G. Teubner, 1927; pp. 83–209.

DOBELLE, W. H. 1977. "Current State of Research on Providing Sight to the Blind by Electrical Stimulation of the Brain." *Journal of Impairment and Blindness* **71**, 290–297.

DUFAU, P. A. 1837. *Essai sur l'état physique, moral et intellectuel des aveugles-nés: avec un nouveau plan pour l'amélioration de leur condition sociale.* Paris: Imprimerie Royale.

DUNAN, C. 1888. "L'espace visuel et l'espace tactile." *Revue philosophique* **25**, 134–169, 354–386, 591–619.

DUTT, K. C. 1938. "Cataract Operations in the Prehistoric Age." *Archives of Ophthalmology* **20**, 1–15.

EBBINGHAUS, H. 1908. *Abriß der Psychologie.* Leipzig: Veit und Co.

ERHARDT-SIEBOLD, E. VON. 1932. "Harmony of the Senses in English, German, and French Romanticism." *Publications of the Modern Language Association of America* **47**, 577–592.

EVANS, G. 1985. "Molyneux's Question." In: Evans, *Collected Papers.* Edited by A. Phillips. Oxford: Clarendon Press; pp. 364–399.

[FONTENELLE, B. LE BOVIER DE]. 1707. "Sur les cataractes des yeux." *Histoire de l'Académie Royale des Sciences* pour l'année 1707, 22–25. Paris: Chez la Veuve de Jean Boudot.

FORREST, D. W. 1965. "Recovery of Sight in the Blind." *American Journal of Psychology* **78**, 147–148.

FORREST, D. W. 1974. "Von Senden, Mesmer, and the Recovery of Sight in the Blind." *American Journal of Psychology* **87**, 719–722.

FRANZ, J. C. A. 1839. *The Eye: A Treatise on the Art of Preserving this Organ in a Healthy Condition, and of Improving the Sight; to which is Prefixed, a View of the Anatomy, and Physiology of the Eye; with Observations on its Expression as Indicative of the Character and Emotions of the Mind.* London: J. Churchill.

FRANZ, J. C. A. 1841. "Memoir of the Case of a Gentleman Born Blind, and Successfully Operated Upon in the 18th Year of his Age, with Physiological Observations and Experiments." *Philosophical Transactions of the Royal Society of London* **131**, 59–68.

GRANT, R. [PSEUD.]. 1709. *A Full and True Account of a Miraculous Cure of a Young Man in Newington, that was Born Blind, and was in Five Minutes brought to Perfect Sight.* By Mr. Roger Grant, Oculist. London.

'S GRAVESANDE, W. J. [1736] 1737. *Introductio ad philosophiam, metaphysicam et logicam continens.* Editio altera. Leidæ: Apud Johannem et Hermannum Verbeek, 1737.

GREGORY, R. L. 1974a. *Concepts and Mechanisms of Perception.* London: Duckworth.

GREGORY, R. L. 1974b. "History of Science: I. The Scientific Past and the Practising Scientist." *Times Literary Supplement*, 26 April, 429–430.

GREGORY, R. L. & WALLACE, J. G. 1963. *Recovery from Early Blindness: A Case Study.* Cambridge: Heffer and Sons.

GRIMM, F. M. & DIDEROT, D. 1813. *Correspondance littéraire, philosophique et critique adressée a un souverain d'Allemagne depuis 1753 jusqu'en 1769.* Paris: Longchamps et F. Buisson.

GUARNIERO, G. 1974. "Experience of Tactile Vision." *Perception* **3**, 101–104.

HAAFF, G. TEN. 1761. *Korte verhandeling, door voorbeelden gesterkt, nopens de nieuwe wyze om de cataracta, met de daardoor veroorzaakte blindheid te genezen, door middel van het kristallyne vocht uit het oog te nemen.* Rotterdam: Reinier Arrenberg.

HAGEN, F. W. 1844. "Psychologie und Psychiatrie." In: Wagner (ed.) 1842–1853, vol. 2 (1844), pp. 692–827.

HAMILTON, W. [1858–1860] 1861–1866. *Lectures on Metaphysics and Logic.* Second revised Edition, edited by Henry Longville Mansel and John Veitch. Edinburgh and London: William Blackwood and Sons, 1861–1866.

HAUSMANN, H. 1989. "William Cheselden (1688–1752): zum 300. Geburtstag des englischen Chirurgen." *Zeitschrift für Urologie und Nephrologie* **82**, 151–154.

HAÜY, V. 1786. *Essai sur l'éducation des aveugles.* Paris: Les Enfans-Aveugles.

HEBB, D. O. 1949. *The Organization of Behavior.* New York: Wiley.

HEIL, J. 1987. "The Molyneux Problem." *Journal for the Theory of Social Behaviour* **27**, 227–241.

HELMHOLTZ, H. L. F. VON. 1856–1867. *Handbuch der physiologischen Optik.* Hamburg und Leipzig: Leopold Voss.

HENKES, H. E. 1982. *Van staarsteker tot oogarts: vier eeuwen oogheelkunde in Rotterdam.* Afscheidsrede, Erasmus Universiteit Rotterdam, 15 September 1982.

HERING, E. [1874] 1878. *Zur Lehre vom Lichtsinne: sechs Mittheilungen.* Zweite unveränderte Abdruck. Wien: Carl Gerolds Sohn.

HIPPEL, A. VON. 1875. "Beobachtungen an einem mit doppelseitiger Cataract geborenen, erfolgreich operirten Kinde." *Von Graefe's Archiv für Ophthalmologie* **21**, 101–131.

HIRSCH, H. V. B. & SPINELLI, D. N. 1970. "Visual Experience Modifies Distribution of Horizontally and Vertically Oriented Receptive Fields in Cats." *Science* **168**, 869–871.

HIRSCHBERG, J. 1875. "Eine Beobachtungsreihe zur empiristischen Theorie des Sehens." *Von Graefe's Archiv für Ophthalmologie* **21**, 23–42.

HIRSCHBERG, J. 1905. "Arabian Ophthalmology." *Journal of the American Medical Association* **45**, 1127–1131.

HOF, M. W. VAN. & WALTER, W. G. [1978] 1984. "Plasticiteit." In: P. E. Voorhoeve (ed.), *Leerboek der neurofysiologie.* Tweede druk. Amsterdam en Brussel: Elsevier, 1984; pp. 521–554.

HOF-VAN DUIN, J. VAN. 1981. "Invloed van zintuiglijke deprivatie op het centrale zenuwstelsel van de kat en van de mens." *Nederlands Tijdschrift voor de Geneeskunde* **125**, 1655–1657.

HOME, E. 1807. "An Account of two Children born with Cataracts in their Eyes, to Shew that their Sight was Obscured in Very Different Degrees: With Experiments to Determine the Proportional Knowledge of Objects Acquired by them Immediately after the Cataracts were Removed." *Philosophical Transactions of the Royal Society of London* **97**, 83–92.

HUME, D. [1748] 1975. *Enquiry into the Human Understanding.* In: Hume, *Enquiries concerning Human Understanding and concerning the Principles of Morals.* Edited by L. A. Selby-Bigge. Third Edition, edited by P. H. Nidditch. Oxford: Clarendon Press, 1975.

NICOLSON, M. H. 1946. *Newton Demands the Muse: Newton's* Opticks *and the Eighteenth Century Poets.* Princeton, N. J.: Princeton University Press.

HOPPEN, K. T. 1963. "The Royal Society and Ireland: William Molyneux F. R. S. (1656–1698)." *Notes and Records of the Royal Society of London* **18**, 125–135.

HOPPEN, K. T. 1970. *The Common Scientist in the Seventeenth Century: A Study of the Dublin Philosophical Society, 1683–1708.* Charlottesville: The University Press of Virginia.

HUBEL, D. H. 1988. *Eye, Brain, and Vision.* New York: Scientific American Library.

HUBEL, D. H. & WIESEL, T. N. 1959. "Receptive Fields of Single Neurones in the Cat's Striate Cortex." *Journal of Physiology* **148**, 574–591.

HUTCHESON, F. [1727] 1788. "Original Letter from Dr. Francis Hutcheson to Mr. William Mace, Professor at Gresham College, 6 September 1727." *European Magazine and London Review*, September 1788, 158–160.

HUTCHESON, F. 1728. *An Essay on the Nature and Conduct of the Passions and Affections: With Illustrations on the Moral Sense.* London: Printed by J. Darby and T. Browne, for John Smith and William Bruce.

HUYGENS, CHR. [1667] 1916. "Deuxième complément à la dioptrique." Reprinted in: Huygens, *Œuvres complètes.* Vol. 13. La Haye: Martinus Nijhoff, 1916.

JACKSON, F. 1982. "Epiphenomenal Qualia." *Philosophical Quarterly* **32**, 127–136.

JAMES, W. 1887. "The Perception of Space." *Mind* **12**, 1–30, 183–211, 321–353, 516–548.

JAMES, W. 1890. *The Principles of Psychology.* New York: Henry Holt and Co.

JANET, P. 1879. "De la perception visuelle de la distance." *Revue philosophique de la France et de l'étranger* **7**, 1–17.

JANIN, J. 1772. *Mémoires et observations anatomiques, physiologiques et physiques sur l'œil, et sur les maladies qui affectent cet organe: avec un précis des opérations et des remèdes qu'on doit pratiquer pour les guérir.* Lyon: Chez les Frêres Perisse, Théophile Barrois; Paris: Didot.

JONES, B. 1975. "Spatial Perception in the Blind." *British Journal of Psychology* **66**, 461–472.

JURIN, J. [PHILALETHES CANTABRIGIENSIS, PSEUD.]. 1734. *Geometry no Friend to Infidelity: Or, a Defence of Sir Isaac Newton and the British Mathematicians, in a Letter to the Author of the Analyst.* By Philalethes Cantabrigiensis. London: Printed for T. Cooper.

JURIN, J. [PHILALETHES CANTABRIGIENSIS, PSEUD.]. 1735. *The Minute Mathematician: Or, the Free-Thinker no Just-Thinker. Set forth in a Second Letter to the Author of the Analyst; Containing a Defence of Sir Isaac Newton and the British Mathematicians against a late Pamphlet, entituled* A Defence of Free-Thinking in Mathematics. By Philalethes Cantabrigiensis. London.

JURIN, J. 1738. "Dr. *Jurin*'s Solution of Mr. *Molyneux*'s Problem." In: R. Smith 1738, vol. 2, pp. 27–29 (remarks upon bk. I, ch. v, art. 132, §§160–170).

KANT, I. [1781] 1787. *Critik der reinen Vernunft.* Zweite Ausgabe. Riga: J. F. Hartknoch, 1787.

KANT, I. 1788. *Critik der practischen Vernunft.* Riga: J. F. Hartknoch.

KAY, L. 1974. "A Sonar Aid to Enhance Spatial Perception of the Blind: Engineering Design and Evaluation." *The Radio and Electronic Engineer* **44**, 605–627.

KELLY, P. 1979. "Locke and Molyneux: The Anatomy of a Friendship." *Hermathena: A Series of Papers on Literature, Science and Philosophy* **126**, 38–54.

KEPLER, J. 1604. *Ad Vitellionem paralipomena, quibus astronomiæ pars optica traditur: potissimum de artificiosa observatione et æstimatione diametrorum deliquiorumque solis et lunæ, cum exemplis insignium eclipsium: habes hoc libro, lector, inter alia multa nova, tractatum luculentum de modo visionis et humorum oculi usu, contra opticos et anatomicos.* Francofurti: Apud Claudium Marnium [etc.].

KLEIN, D. B. 1970. *A History of Scientific Psychology: Its Origins and Philosophical Backgrounds.* London: Routledge and Kegan Paul.

KÖHLER, W. 1947. *Gestalt Psychology: An Introduction to New Concepts in Modern Psychology.* New York: Liveright Publishing Corporation.

KRÜGER, J. G. 1756. *Versuch einer Experimentalseelenlehre.* Halle und Helmstedt: Hemmerde.

LA METTRIE, J. OFFROY DE. 1745. *Histoire naturelle de l'âme, traduite de l'anglois de M. Charp, par feu M. H** de l'Académie des Sciences, &c.* La Haye: Jean Néaulme.

LA METTRIE, J. OFFROY DE. 1748. *L'homme machine.* Leyde: Imprimerie de E. Luzac fils.

LA METTRIE, J. OFFROY DE. 1751. *Traité de l'âme.* In: La Mettrie, *Œuvres philosophiques.* Londres: Jean Nourse; pp. 81–208.

LEBENSOHN, J. E. (ED.) 1969. *An Anthology of Ophthalmic Classics*. Baltimore: The Williams and Wilkins Co.

LE CAT, C. N. [1740] 1744. *Traité des sens*. Rouen 1740. Nouvelle édition, corrigée, augmentée et enrichie de figures en taille douce. Amsterdam: J. Wetstein, 1744.

LEE, H. 1702. *Anti-Scepticism: Or, Notes Upon each Chapter of Mr. Lock's Essay concerning Humane Understanding, With an Explication of All the Particulars of which he Treats, and in the same Order*. London: R. Clavel and C. Harper.

LEIBNIZ, G. W. [1765] 1966. *Nouveaux essais sur l'entendement humain*. Amsterdam et Leipzig: Rudolf Erich Raspe, 1765. Chronologie et introduction par Jacques Brunschwig. Paris: Garnier-Flammarion, 1966.

LENDE, H. [1940] 1953. *Books about the Blind: A Bibliographical Guide to Literature Relating to the Blind*. New revised Edition. New York: American Foundation for the Blind, 1953.

LEVIN, J. 1986. "Could Love be Like a Heatwave? Physicalism and the Subjective Character of Experience." *Philosophical Studies* **49**, 245–261.

LIEVERS, M. 1992. "The Molyneux Problem." *Journal of the History of Philosophy* **30**, 399–416.

LOCKE, J. 1688. "Extrait d'un livre anglois qui n'est pas encore publié, intitulé *Essai philosophique concernant l'entendement*, où l'on montre quelle est l'étenduë de nos connoissances certaines, et la manière dont nous y parvenons." *Bibliothèque universelle et historique* **8**, 49–142. Amsterdam: Chez Wolfgang, Waesberge, Boom, et Van Someren.

LOCKE, J. [1690] 1975. *An Essay concerning Humane Understanding*. London: Printed by Eliz. Holt for Thomas Basset. Reprinted in: Locke 1975.

LOCKE, J. [1694] 1975. *An Essay concerning Humane Understanding*. Second Edition. London: Printed for Awnsham and John Churchil, and Samuel Manship. Reprinted in: Locke 1975.

LOCKE, J. 1700. *Essai philosophique concernant l'entendement humain*. Traduit de l'anglois par Pierre Coste, sur la quatrième édition. Amsterdam: H. Schelte.

LOCKE, J. 1708. *Some Familiar Letters between Mr. Locke and Several of his Friends*. London: A. and J. Churchill.

LOCKE, J. 1975. *An Essay concerning Human Understanding*. Edited with a Foreword by Peter H. Nidditch. Oxford: Clarendon Press.

LOCKE, J. 1976–1990. *The Correspondence of John Locke*. Edited by E. S. de Beer. Oxford: Clarendon Press.

LOTZE, H. [1881] 1882. *Grundzüge der Psychologie: Dictate aus den Vorlesungen von Hermann Lotze*. Zweite Auflage. Leipzig: S. Hirzel, 1882.

LOUIS, A. 1752. "Cataracte." In: Diderot & D'Alembert (eds.) 1751–1780, vol. 2 (1752), pp. 770–771.

LUCE, A. A. [1934] 1967. *Berkeley and Malebranche: A Study in the Origins of Berkeley's Thought*. Second Edition. Oxford: Clarendon Press, 1967.

MACH, E. 1886. *Beiträge zur Analyse der Empfindungen*. Jena: Gustav Fischer.

MACH, E. 1900. *Die Analyse der Empfindungen und das Verhältniss des Physischen zum Psychischen.* Zweite vermehrte Auflage der *Beiträge zur Analyse der Empfindungen.* Jena: Gustav Fischer.

MACKIE, J. L. 1974. "Philosophical Prediction." *Times Literary Supplement*, 3 May, 477.

MACKIE, J. L. 1976. *Problems from Locke.* Oxford: Oxford University Press.

MACLEAN, K. 1936. *John Locke and English Literature of the Eighteenth Century.* New Haven: Yale University Press.

MAÎTRE-JAN, A. 1707. *Traité des maladies de l'œil et des remèdes propres pour leur guérison.* Troyes: Le Febvre.

MARCHAN, L. 1768. *Observatien over de quaalen der oogen, en 't gebruik van een ophtalmickswaater, om die te geneezen.* Uit het Fransch. 's-Gravenhage: D. Monnier.

MÉRIAN, J. B. [1772/73/74/76/77/79/81/82] 1984. "Sur le problème de Molyneux: premier–huitiéme mémoire." *Nouveaux mémoires de l'Académie Royale des Sciences et Belles-Lettres* pour l'année 1770/71/72/74/75/77/79/80. Berlin: Voss, 1772/73/74/76/77; Berlin: Decker, 1779/81/82. Reprinted in: Mérian, *Sur le problème de Molyneux.* Suivi de "Mérian, Diderot et l'Aveugle" par Francine Markovits. Paris: Flammarion.

MILL, J. S. [1842] 1859. "Bailey on Berkeley's Theory of Vision." *Westminster Review*, October 1842. Reprinted in: Mill 1859, pp. 84–114.

MILL, J. S. [1843] 1859. "Rejoinder to Mr. Bailey's Reply." *Westminster Review*, May 1843. Reprinted in: Mill 1859, pp. 114–119.

MILL, J. S. 1859. *Dissertations and Discussions: Political, Philosophical, and Historical.* Vol. 2. London: Longmans, Green, Reader and Dyer.

MILL, J. S. 1865. *An Examination of Sir William Hamilton's Philosophy and of the Principal Philosophical Questions Discussed in his Writings.* London: Longman, Green, Longman, Roberts and Green.

MOLYNEUX, C. 1820. *An Account of the Family and Descendants of Sir Thomas Molyneux, Kt., Chancellor of the Exchequer in Ireland to Queen Elizabeth.* Evesham: J. Agg.

MOLYNEUX, W. 1692. *Dioptrica Nova: A Treatise of Dioptricks, in Two Parts, Wherein the Various Effects and Appearances of Spherick Glasses, Both Convex and Concave, Single and Combined, in Telescopes and Microscopes, Together with their Usefulness in many Concerns of Humane Life, are Explained.* London: Printed for Benj. Tooke.

MOLYNEUX, W. 1698. *The Case of Ireland's Being Bound by Acts of Parliament in England, Stated.* Dublin: Joseph Ray.

MOLYNEUX, W. 1803. *Anecdotes of the Life of the Celebrated Patriot and Philosopher Wm. Molyneux, Author of* The Case of Ireland. Published from a Manuscript Written by Himself, in Possession of Sir Capel Molyneux, to Which are Annexed Explanatory Notes by the Editor. Dublin: Privately printed.

MONCK, W. H. S. 1882. "Observations in Cases of Couching for Cataract." *Mind* **7**, 107–110.

MORAND, S. F. 1757. "Éloge de M. Cheselden." *Mémoires de l'Académie Royale de Chirurgie* **7**, 168–190.

MORGAN, M. J. 1977. *Molyneux's Question: Vision, Touch and the Philosophy of Perception.* Cambridge: Cambridge University Press.

MÜLLER, J. 1826. *Zur vergleichenden Physiologie des Gesichtssinnes des Menschen und der Thiere nebst ein Versuch über die Bewegungen der Augen und über den menschlichen Blick.* Leipzig: C. Cnobloch.

MÜLLER, J. 1837–1840. *Handbuch der Physiologie des Menschen für Vorlesungen.* Coblenz: J. Hölscher.

MÜNCHOW, W. 1984. *Geschichte der Augenheilkunde.* Separatdruck aus "Der Augenarzt," Band 9. Zweite ergänzte und überarbeitete Auflage. Stuttgart: Ferdinand Enke.

MURRAY, D. J. 1983. *A History of Western Psychology.* Englewood Cliffs, N. J.: Prentice-Hall.

MURRAY, M. 1944. "An Introduction to Bishop Berkeley's Theory of Vision." *The British Journal of Ophthalmology* **28**, 600–611.

MUSSCHENBROEK, P. VAN. 1762. *Introductio ad philosophiam naturalem.* Lugduni Batavorum: Sam. et Joh. Luchtmans.

NEWTON, I. 1704. *Opticks: Or a Treatise of the Reflections, Refractions, Inflections and Colours of Light.* London: Sam. Smith and Benj. Walford.

NUNNELEY, T. 1858. *The Organs of Vision: Their Anatomy and Physiology.* London: J. Churchill.

PANUM, P. L. 1858. *Physiologische Untersuchungen über das Sehen mit zwei Augen.* Kiel: Schwerssche Buchhandlung.

PARK, D. 1969. "Locke and Berkeley on the Molyneux Problem." *Journal of the History of Ideas* **30**, 253–260.

PASTORE, N. 1971. *A Selective History of Theories of Visual Perception: 1650–1950.* New York: Oxford University Press.

PASTORE, N. 1973. "On Plagiarism: Buffon, Condillac, Porterfield, Schopenhauer." *Journal of the History of Behavioral Sciences* **9**, 378–392.

PATER, C. DE. 1979. *Petrus van Musschenbroek (1692–1761): een Newtoniaans natuuronderzoeker.* Proefschrift. Utrecht: Elinkwijk.

PATER, C. DE. 1988. *Willem Jacob 's Gravesande: welzijn, wijsbegeerte en wetenschap.* Uitgegeven, ingeleid en van aantekeningen voorzien door C. de Pater. Baarn: Ambo.

PAULSON, W. R. 1987. *Enlightenment, Romanticism, and the Blind in France.* Princeton, N. J.: Princeton University Press.

PERKINS, M. L. 1978. "The Crisis of Sensationalism in Diderot's *Lettre sur les aveugles.*" *Studies on Voltaire and the Eighteenth Century* **174**, 167–188.

PITCHER, G. 1974. "Thomson's Problem." *The Journal of Philosophy* **71**, 651–652.

PLATNER, E. [1776] 1793. *Philosophische Aphorismen nebst einigen Anleitungen zur philosophischen Geschichte.* Ganz neue Ausarbeitung. Leipzig: Schwickert, 1793.

PORTERFIELD, W. 1759. *A Treatise on the Eye, the Manner and Phænomena of Vision.* Edinburgh: G. Hamilton and J. Balfour.

PREYER, W. 1882. *Die Seele des Kindes: Beobachtungen über die geistige Entwicklung des Menschen in den ersten Lebensjahren.* Leipzig: Th. Grieben.

PREYER, W. 1908. *Die Seele des Kindes: Beobachtungen über die geistige Entwicklung des Menschen in den ersten Lebensjahren.* Siebente Auflage, bearbeitet und herausgegeben von K. L. Schaefer. Leipzig: Th. Grieben.

PRIESTLEY, J. 1772. *The History and Present State of Discoveries Relating to Vision, Light and Colours.* London: J. Johnson.

RAEHLMANN, E. 1891. "Physiologisch-psychologische Studien über die Entwicklung der Gesichtswahrnehmungen bei Kindern und bei operierten Blindgeborenen." *Zeitschrift für Psychologie und Physiologie der Sinnesorgane* **2**, 53–96.

RAMSAY, A. M. 1903. "Case of a Man Blind from Congenital Cataract who Acquired Sight after an Operation when he was 30 Years of Age." *The Lancet*, 16 May, 1364–1366.

REID, T. [1764] 1970. *An Inquiry into the Human Mind.* Edited with an Introduction by T. Duggan. Chicago and London: University of Chicago Press, 1970.

REID, T. [1837] 1863. *The Works of Thomas Reid, D. D.* Preface, Notes and Supplementary Dissertations by Sir W. Hamilton. Sixth edition. Edinburgh: MacLachlan and Stewart; London: Longman, Green, Longman, Roberts, and Green, 1863.

RIESEN, A. H. 1947. "The Development of Visual Perception in Man and Chimpanzee." *Science* **106**, 107–108.

RIESEN, A. H. 1950. "Arrested Vision." *Scientific American* **183**, 16–19.

RIESEN, A. H. 1960. "Effects of Stimulus Deprivation on the Development and Atrophy of the Visual Sensory System." *American Journal of Orthopsychiatry* **30**, 23–36.

ROCHAT, G. F. 1939. "Petrus Camper als oogheelkundige." *Nederlands Tijdschrift voor Geneeskunde* **83**, 2101–2106.

ROHAULT, J. 1672. *Traité de physique.* Amsterdam: Jacques le Jeune.

SABRA, A. I. 1967. *Theories of Light: From Descartes to Newton.* Cambridge: Cambridge University Press.

SACKS, O. 1993. "To See and Not See." *New Yorker* **69**, 59–73.

SAUNDERSON, N. 1740. *The Elements of Algebra.* Cambridge: The University Press.

SCHLODTMANN, W. 1902. "Ein Beitrag zur Lehre von der optischen Localisation bei Blindgeborenen." *Von Graefe's Archiv für Ophthalmologie* **54**, 256–267.

SENDEN, M. VON. [1932] 1960. *Raum- und Gestaltauffassung bei operierten Blindgeborenen.* Leipzig: J. A. Barth, 1932. English translation: *Space and Sight: The Perception of Space and Shape in the Congenitally Blind before and after Operation.* Translated by P. Heath. With Appendixes by A. H. Riesen, G. J. Warnock and J. Z. Young. London: Methuen, 1960.

SIMMS, J. G. 1982. *William Molyneux of Dublin: A Life of the Seventeenth-Century Political Writer and Scientist.* Edited by P. H. Kelly. Blackrock: Irish Academic Press.

SMITH, A. 1795. *Essays on Philosophical Subjects.* Dublin: Printed for Messrs. Wogan, Byrne, J. Moore, Colbert, Rice, W. Jones, Porter, and Folingsby.

SMITH, R. 1738. *A Compleat System of Opticks in Four Books, viz. a Popular, a Mathematical, a Mechanical, and a Philosophical Treatise: To which are Added Remarks upon the Whole.* Cambridge: Printed for the Author, and sold there by Cornelius Crownfield, and at London by Stephen Austen, and Robert Dodsley.

SPALDING, D. A. 1873. "Instinct: With original Observations on young Animals." *Macmillan's Magazine* **27**, 282–293.

[STEELE, R.]. [1709] 1898. [An account of Mr. Grant's operation of 29 June 1709.] *The Tatler* **1**, No. 55, Tuesday 16 August 1709. Reprinted in: *The Tatler.* Edited by G. A. Aitken. Vol. 2. London: Duckworth, 1898; pp. 41–46.

STEWART, D. [1815] 1854. "Some Account of a Boy born Blind and Deaf, Collected from Authentic Sources of Information: With a few Remarks and Comments." *Transactions of the Royal Society of Edinburgh* **7**, 1–78. Reprinted with Supplement (1827) in: Stewart, "Appendix to Part III, Ch. 2, of the Second Edition of *Elements of the Philosophy of the Human Mind* (1827)." Reprinted in: *The Collected Works of Dugald Stewart.* Edited by Sir W. Hamilton. Vol. 4. Edinburgh: T. and T. Clark, 1854; pp. 300–370; with "Appendix by Editor" on pp. 388–389.

STRATTON, G. M. 1899. "The Spatial Harmony of Sight and Touch." *Mind* **24**, 492–505.

STUMPF, C. 1873. *Über den psychologischen Ursprung der Raumvorstellung.* Leipzig: S. Hirzel.

SYNGE, E. 1693. *A Gentleman's Religion.* London: Printed for A. and J. Churchill, and R. Sare.

TEAPE, C. R. 1870. *Berkeleian Philosophy.* Edinburgh: Colston and Son.

THOMSON, J. J. 1974. "Molyneux's Problem." *The Journal of Philosophy* **71**, 637–650.

TOURTUAL, C. T. 1827. *Die Sinne des Menschen in den wechselseitigen Beziehungen ihres psychischen und organischen Lebens: ein Beitrag zur physiologischen Æsthetik.* Münster: Regensburg.

TREVOR-ROPER, P. D. & CURRAN, P. V. [1974] 1984. *The Eye and its Disorders.* Second Edition. Oxford: Blackwell Scientific Publications, 1984.

[TRÉVOUX]. 1714. *Index funereus chirurgorum Parisiensium ab anno 1315 ad annum 1714.* Paris: Ganeau.

TURBAYNE, C. M. 1955. "Berkeley and Molyneux on Retinal Images." *Journal of the History of Ideas* **16**, 339–355.

VALVO, A. 1971. *Sight Restoration after Long Term Blindness: Two Problems and Behaviour Patterns of Visual Rehabilitation.* New York: American Foundation for the Blind.

VERBEEK, T. H. M. 1988. *Le* Traité de l'âme *de La Mettrie: édition critique du texte avec une introduction et un commentaire historiques.* Utrecht: OMI-Grafisch Bedrijf.

VIENNE, J.-M. 1992. "Locke et l'intentionnalité: le problème de Molyneux." *Archives de philosophie* **55**, 661–684.

VILLEY, P. 1914. *Le monde des aveugles: essai de psychologie.* Paris: Ernest Flammarion.

VOLKMANN, A. W. 1846. "Sehen." In: Wagner, (ed.) 1842–1853, vol. 3 (1846), pp. 265–351.

VOLKMANN, A. W. 1863. *Physiologische Untersuchungen im Gebiete der Optik.* Leipzig: Breitkopf und Härtel.

VOLTAIRE. [1738] 1819. *Elémens de la philosophie de Neuton.* Amsterdam: Jacques Desbordes, 1738. Reprinted in: *Œuvres complètes de Voltaire.* (Édition de Kehl.) Vol. 28. Paris: Antoine-Augustin Renouard, 1819.

WAGNER, R. (ED.) 1842–1853. *Handwörterbuch der Physiologie mit Rücksicht auf physiologische Pathologie.* Braunschweig: Friedr. Vieweg und Sohn.

WARDROP, J. 1826. "Case of a Lady Born Blind, who Received Sight at an Advanced Age by the Formation of an Artificial Pupil." *Philosophical Transactions of the Royal Society of London* **116**, 529–540.

WARE, J. 1801. "Case of a Young Gentleman, who Recovered his Sight when Seven Years of Age, after having been Deprived of it by Cataracts, before he was a Year Old: With Remarks." *Philosophical Transactions of the Royal Society of London* **91**, 382–396.

WARREN, D. H. & STRELOW, E. R. 1984. "Learning Spatial Dimensions with a Visual Sensory Aid: Molyneux Revisited." *Perception* **13**, 331–350.

WHEATSTONE, C. 1838. "Contributions to the Physiology of Vision—Part the First: On some Remarkable, and Hitherto Unobserved, Phenomena of Binocular Vision." *Philosophical Transactions of the Royal Society of London* **128**, 371–394.

WHITE, B. W., SAUNDERS, F. A., SCADDEN, L., BACH-Y-RITA, P. & COLLINS, C. C. 1970. "Seeing with the Skin." *Perception and Psychophysics* **7**, 23–27.

WILKES, K. V. 1988. *Real People: Personal Identity Without Thought Experiments.* Oxford: Clarendon Press.

WUNDT, W. [1896] 1907. *Grundriß der Psychologie.* Achte, verbesserte Auflage. Leipzig: Wilhelm Engelmann, 1907.

ZEKI, S. 1993. *A Vision of the Brain.* Oxford: Blackwell Scientific Publications.

ZEUNE, A. 1821. *Belisar: über Blinde und Blinden-Anstalten.* Berlin: In der Blinden-Anstalt.

ZUCKERMAN, C. B. & ROCK, I. 1957. "A Reappraisal of the Rôles of Past Experience and Innate Organizing Processes in Visual Perception." *Psychological Bulletin* **54**, 269–296.

INDEX OF NAMES

Page numbers in italics refer to the bibliography.

INDEX OF SUBJECTS

ARCHIVES INTERNATIONALES D'HISTOIRE DES IDÉES
*
INTERNATIONAL ARCHIVES OF THE HISTORY OF IDEAS

85. Bérault Stuart, Seigneur d'Aubigny: *Traité sur l'art de la guerre.* Introduction et édition par Élie de Comminges. 1976 ISBN 90-247-1871-6
86. S.L. Kaplan: *Bread, Politics and Political Economy in the Reign of Louis XV.* 2 vols., 1976 Set ISBN 90-247-1873-2
87. M. Lienhard (ed.): *The Origins and Characteristics of Anabaptism / Les débuts et les caractéristiques de l'Anabaptisme.* With an Extensive Bibliography / Avec une bibliographie détaillée. 1977 ISBN 90-247-1896-1
88. R. Descartes: *Règles utiles et claires pour la direction de l'esprit en la recherche de la vérité.* Traduction selon le lexique cartésien, et annotation conceptuelle par J.-L. Marion. Avec des notes mathématiques de P. Costabel. 1977 ISBN 90-247-1907-0
89. K. Hardesty: *The 'Supplément' to the 'Encyclopédie'.* [Diderot et d'Alembert]. 1977 ISBN 90-247-1965-8
90. H.B. White: *Antiquity Forgot.* Essays on Shakespeare, [Francis] Bacon, and Rembrandt. 1978 ISBN 90-247-1971-2
91. P.B.M. Blaas: *Continuity and Anachronism.* Parliamentary and Constitutional Development in Whig Historiography and in the Anti-Whig Reaction between 1890 and 1930. 1978 ISBN 90-247-2063-X
92. S.L. Kaplan (ed.): *La Bagarre.* Ferdinando Galiani's (1728-1787) 'Lost' Parody. With an Introduction by the Editor. 1979 ISBN 90-247-2125-3
93. E. McNiven Hine: *A Critical Study of [Étienne Bonnot de] Condillac's* [1714-1780] *'Traité des Systèmes'.* 1979 ISBN 90-247-2120-2
94. M.R.G. Spiller: *Concerning Natural Experimental Philosphy.* Meric Casaubon [1599-1671] and the Royal Society. 1980 ISBN 90-247-2414-7
95. F. Duchesneau: *La physiologie des Lumières.* Empirisme, modèles et théories. 1982 ISBN 90-247-2500-3
96. M. Heyd: *Between Orthodoxy and the Enlightenment.* Jean-Robert Chouet [1642-1731] and the Introduction of Cartesian Science in the Academy of Geneva. 1982 ISBN 90-247-2508-9
97. James O'Higgins: *Yves de Vallone* [1666/7-1705]: *The Making of an Esprit Fort.* 1982 ISBN 90-247-2520-8
98. M.L. Kuntz: *Guillaume Postel* [1510-1581]. Prophet of the Restitution of All Things. His Life and Thought. 1981 ISBN 90-247-2523-2
99. A. Rosenberg: *Nicolas Gueudeville and His Work (1652-172?).* 1982 ISBN 90-247-2533-X
100. S.L. Jaki: *Uneasy Genius: The Life and Work of Pierre Duhem* [1861-1916]. 1984 ISBN 90-247-2897-5; Pb (1987) 90-247-3532-7
101. Anne Conway [1631-1679]: *The Principles of the Most Ancient Modern Philosophy.* Edited and with an Introduction by P. Loptson. 1982 ISBN 90-247-2671-9
102. E.C. Patterson: *[Mrs.] Mary [Fairfax Greig] Sommerville* [1780-1872] *and the Cultivation of Science (1815-1840).* 1983 ISBN 90-247-2823-1
103. C.J. Berry: *Hume, Hegel and Human Nature.* 1982 ISBN 90-247-2682-4
104. C.J. Betts: *Early Deism in France.* From the so-called 'déistes' of Lyon (1564) to Voltaire's 'Lettres philosophiques' (1734). 1984 ISBN 90-247-2923-8

ARCHIVES INTERNATIONALES D'HISTOIRE DES IDÉES
*
INTERNATIONAL ARCHIVES OF THE HISTORY OF IDEAS

105. R. Gascoigne: *Religion, Rationality and Community.* Sacred and Secular in the Thought of Hegel and His Critics. 1985 ISBN 90-247-2992-0
106. S. Tweyman: *Scepticism and Belief in Hume's 'Dialogues Concerning Natural Religion'*. 1986 ISBN 90-247-3090-2
107. G. Cerny: *Theology, Politics and Letters at the Crossroads of European Civilization.* Jacques Basnage [1653-1723] and the Baylean Huguenot Refugees in the Dutch Republic. 1987 ISBN 90-247-3150-X
108. Spinoza's *Algebraic Calculation of the Rainbow & Calculation of Changes.* Edited and Translated from Dutch, with an Introduction, Explanatory Notes and an Appendix by M.J. Petry. 1985 ISBN 90-247-3149-6
109. R.G. McRae: *Philosophy and the Absolute.* The Modes of Hegel's Speculation. 1985 ISBN 90-247-3151-8
110. J.D. North and J.J. Roche (eds.): *The Light of Nature.* Essays in the History and Philosophy of Science presented to A.C. Crombie. 1985 ISBN 90-247-3165-8
111. C. Walton and P.J. Johnson (eds.): *[Thomas] Hobbes's 'Science of Natural Justice'.* 1987 ISBN 90-247-3226-3
112. B.W. Head: *Ideology and Social Science.* Destutt de Tracy and French Liberalism. 1985 ISBN 90-247-3228-X
113. A.Th. Peperzak: *Philosophy and Politics.* A Commentary on the Preface to Hegel's *Philosophy of Right.* 1987 ISBN Hb 90-247-3337-5; Pb ISBN 90-247-3338-3
114. S. Pines and Y. Yovel (eds.): *Maimonides* [1135-1204] *and Philosophy.* Papers Presented at the 6th Jerusalem Philosophical Encounter (May 1985). 1986 ISBN 90-247-3439-8
115. T.J. Saxby: *The Quest for the New Jerusalem, Jean de Labadie* [1610-1674] *and the Labadists (1610-1744).* 1987 ISBN 90-247-3485-1
116. C.E. Harline: *Pamphlets, Printing, and Political Culture in the Early Dutch Republic.* 1987 ISBN 90-247-3511-4
117. R.A. Watson and J.E. Force (eds.): *The Sceptical Mode in Modern Philosophy.* Essays in Honor of Richard H. Popkin. 1988 ISBN 90-247-3584-X
118. R.T. Bienvenu and M. Feingold (eds.): *In the Presence of the Past.* Essays in Honor of Frank Manuel. 1991 ISBN 0-7923-1008-X
119. J. van den Berg and E.G.E. van der Wall (eds.): *Jewish-Christian Relations in the 17th Century.* Studies and Documents. 1988 ISBN 90-247-3617-X
120. N. Waszek: *The Scottish Enlightenment and Hegel's Account of 'Civil Society'.* 1988 ISBN 90-247-3596-3
121. J. Walker (ed.): *Thought and Faith in the Philosophy of Hegel.* 1991 ISBN 0-7923-1234-1
122. Henry More [1614-1687]: *The Immortality of the Soul.* Edited with Introduction and Notes by A. Jacob. 1987 ISBN 90-247-3512-2
123. P.B. Scheurer and G. Debrock (eds.): *Newton's Scientific and Philosophical Legacy.* 1988 ISBN 90-247-3723-0
124. D.R. Kelley and R.H. Popkin (eds.): *The Shapes of Knowledge from the Renaissance to the Enlightenment.* 1991 ISBN 0-7923-1259-7

ARCHIVES INTERNATIONALES D'HISTOIRE DES IDÉES
*
INTERNATIONAL ARCHIVES OF THE HISTORY OF IDEAS

---

125. R.M. Golden (ed.): *The Huguenot Connection.* The Edict of Nantes, Its Revocation, and Early French Migration to South Carolina. 1988 ISBN 90-247-3645-5
126. S. Lindroth: *Les chemins du savoir en Suède.* De la fondation de l'Université d'Upsal à Jacob Berzelius. Études et Portraits. Traduit du suédois, présenté et annoté par J.-F. Battail. Avec une introduction sur Sten Lindroth par G. Eriksson. 1988 ISBN 90-247-3579-3
127. S. Hutton (ed.): *Henry More (1614-1687). Tercentenary Studies.* With a Biography and Bibliography by R. Crocker. 1989 ISBN 0-7923-0095-5
128. Y. Yovel (ed.): *Kant's Practical Philosophy Reconsidered.* Papers Presented at the 7th Jerusalem Philosophical Encounter (December 1986). 1989 ISBN 0-7923-0405-5
129. J.E. Force and R.H. Popkin: *Essays on the Context, Nature, and Influence of Isaac Newton's Theology.* 1990 ISBN 0-7923-0583-3
130. N. Capaldi and D.W. Livingston (eds.): *Liberty in Hume's 'History of England'.* 1990 ISBN 0-7923-0650-3
131. W. Brand: *Hume's Theory of Moral Judgment.* A Study in the Unity of *A Treatise of Human Nature.* 1992 ISBN 0-7923-1415-8
132. C.E. Harline (ed.): *The Rhyme and Reason of Politics in Early Modern Europe.* Collected Essays of Herbert H. Rowen. 1992 ISBN 0-7923-1527-8
133. N. Malebranche: *Treatise on Ethics* (1684). Translated and edited by C. Walton. 1993 ISBN 0-7923-1763-7
134. B.C. Southgate: *'Covetous of Truth'.* The Life and Work of Thomas White (1593–1676). 1993 ISBN 0-7923-1926-5
135. G. Santinello, C.W.T. Blackwell and Ph. Weller (eds.): *Models of the History of Philosophy.* Vol. 1: From its Origins in the Renaissance to the 'Historia Philosphica'. 1993 ISBN 0-7923-2200-2
136. M.J. Petry (ed.): *Hegel and Newtonianism.* 1993 ISBN 0-7923-2202-9
137. Otto von Guericke: *The New (so-called Magdeburg) Experiments* [Experimenta Nova, Amsterdam 1672]. Translated and edited by M.G.Foley Ames. 1994 ISBN 0-7923-2399-8
138. R.H. Popkin and G.M. Weiner (eds.): *Jewish Christians and Cristian Jews.* From the Renaissance to the Enlightenment. 1994 ISBN 0-7923-2452-8
139. J.E. Force and R.H. Popkin (eds.): *The Books of Nature and Scripture.* Recent Essays on Natural Philosophy, Theology, and Biblical Criticism in the Netherlands of Spinoza's Time and the British Isles of Newton's Time. 1994 ISBN 0-7923-2467-6
140. P. Rattansi and A. Clericuzio (eds.): *Alchemy and Chemistry in the 16th and 17th Centuries.* 1994 ISBN 0-7923-2573-7
141. S. Jayne: *Plato in Renaissance England.* 1995 ISBN 0-7923-3060-9
142. A.P. Coudert: *Leibniz and the Kabbalah.* 1995 ISBN 0-7923-3114-1
143. M.H. Hoffheimer: *Eduard Gans and the Hegelian Philosophy of Law.* 1995 ISBN 0-7923-3114-1
144. J.R.M. Neto: *The Christianization of Pyrrhonism.* Scepticism and Faith in Pascal, Kierkegaard, and Shestov. 1995 ISBN 0-7923-3381-0
145. R.H. Popkin (ed.): *Scepticism in the History of Philosophy.* A Pan-American Dialogue. 1996 ISBN 0-7923-3769-7

ARCHIVES INTERNATIONALES D'HISTOIRE DES IDÉES

*

INTERNATIONAL ARCHIVES OF THE HISTORY OF IDEAS

---

146. M. de Baar, M. Löwensteyn, M. Monteiro and A.A. Sneller (eds.): *Choosing the Better Part.* Anna Maria van Schurman (1607–1678). 1996 ISBN 0-7923-3799-9
147. M. Degenaar: *Molyneux's Problem.* Three Centuries of Discussion on the Perception of Forms. 1996 ISBN 0-7923-3934-7

KLUWER ACADEMIC PUBLISHERS – DORDRECHT / BOSTON / LONDON

GPSR Compliance

The European Union's (EU) General Product Safety Regulation (GPSR) is a set of rules that requires consumer products to be safe and our obligations to ensure this.

If you have any concerns about our products, you can contact us on ProductSafety@springernature.com

In case Publisher is established outside the EU, the EU authorized representative is:

Springer Nature Customer Service Center GmbH
Europaplatz 3
69115 Heidelberg, Germany

Batch number: 10823959

Printed by Printforce, the Netherlands